Classification and Indexing in Science

Classification and Indexing in Science

B. C. VICKERY
MA, FLA

Professor of Library Studies in the University of London

Third Edition

LONDON
BUTTERWORTHS

THE BUTTERWORTH GROUP

ENGLAND
Butterworth & Co (Publishers) Ltd
London: 88 Kingsway, WC2B 6AB

AUSTRALIA
Butterworths Pty Ltd
Sydney: 586 Pacific Highway, NSW 2067
Melbourne: 343 Little Collins Street, 3000
Brisbane: 240 Queen Street, 4000

CANADA
Butterworth & Co (Canada) Ltd
Toronto: 2265 Midland Avenue, Scarborough MIP 451

NEW ZEALAND
Butterworths of New Zealand Ltd
Wellington: 26–28 Waring Taylor Street, 1

SOUTH AFRICA
Butterworth & Co (South Africa) (Pty) Ltd
Durban: 152–154 Gale Street

First Published, 1958
Second Edition (Enlarged), 1958
Third Edition (Completely Revised), 1975

ISBN 0 408 70662 7

Typeset and printed in Great Britain by
REDWOOD BURN LTD
Trowbridge & Esher

PREFACE TO THE THIRD EDITION

The preface to the first edition of this book—which is reproduced following this—shows that in 1958 the classification ideas in it were felt to be controversial, needing to be championed. A few years before, the Classification Research Group had issued a memorandum proclaiming 'the need for a faceted classification as the basis of all methods of information retrieval'. As part-author of this memorandum, I must now judge the claim to have been too bold, even brash. But it has been vindicated to an extent, for both in theory and in practice the value of facet analysis, in the organisation of subject vocabularies for indexing and search, has been widely accepted—whether these vocabularies are classified or alphabetical, and whether used in pre- or post-coordinate fashion.

This acceptance would in itself urge the adoption of a less polemical and more expository tone in the text. Moreover, the information scene and its use of classification has considerably altered. The result is that—with the exception of the chapter on notation—the main text has been largely rewritten. Some amendments have been made to the appendixes.

During the last 7 years my work with Aslib Research and Development Department gave me the opportunity to study the classification and indexing problems of a number of scientific and technical institutions, and this experience is reflected in some of the views now put forward. I am grateful to my former Aslib colleagues for much fruitful interaction. Although it is now many years since I was an attending member of the Classification Research Group, I have continued to follow their work with interest, and the acknowledgements to individuals in the 1958 preface need to be repeated here.

Acknowledgements are also due to the Library Association for permission to reproduce in Appendix D an essay that was originally published in their *Sayers Memorial Volume*.

B. C. Vickery

PREFACE TO THE FIRST EDITION

'In science', commented Lord Rayleigh in 1884, 'by a fiction as remarkable as any to be found in law, what has once been published, even though it be in the Russian language, is spoken of as *known*, and it is too often forgotten that the rediscovery in the library may be a more difficult and uncertain process than the first discovery in the laboratory.' These words are far truer today than when they were first written, and in 1948 the editors of a Royal Society conference declared that 'the task of keeping up with scientific literature is becoming an impossible one and is in turn leading to inefficiency and to a certain amount of frustration in scientific research and in the application of science'.

This grave situation is giving rise to new ideas and experiments in every phase of scientific communication. The problems are partly due to the sheer bulk of scientific literature, but there are many other aspects. Publications are not only numberless but also of very varied physical and bibliographical form—books, papers, bulletins, theses, private reports, patents, and so on—and of very varied origin and language. To learn of their existence, to acquire them, to make them usable by the reader—each step presents a host of difficulties.

The first problem, that of learning of a publication's existence, is tackled by a multiplicity of abstracting and indexing journals and other bibliographies, and, at local levels, by library catalogues and indexing systems. The administrative difficulties of cutting out unnecessary overlapping among these services, while ensuring adequately comprehensive coverage, are very great. But even if these were overcome, problems of the internal arrangement of these bibliographical aids would still remain. It is not enough for them collectively to record every scientific publication. The user must be able to *find* every such record, starting only with a subject on which he wants information.

INFORMATION RETRIEVAL

It is for this reason that the problem of subject indexing—or, in more general terms, that of information retrieval—assumes such importance in science today. A small index presents few problems, but the analysis by subject of a large bibliography of thousands or even millions of entries is a highly intricate procedure, which cannot be carried out by rule-of-thumb methods.

It is little wonder that for the last 10 years or so there has been a flood of papers, books, lectures, symposia and discussions on indexing, classification, mechanical selection, subject analysis, information retrieval, and so on. We have been showered with new terms for new techniques—and new terms for old ones. An earlier generation used subject headings; today we use facets and phases, isolates and analets, descriptors and uniterms, and many new 'retrieval systems' are proposed.

For too long the features which distinguish one particular system from others have been emphasised at the expense of principles common to all systems. This is particularly true at the present time. Alphabetical indexing, classification and mechanical selection have been made to appear completely different techniques, and are described in such different terms that workers in one field are sometimes unintelligible to those in others. The first purpose of this book is, therefore, to describe trends which reveal that all methods of subject indexing face similar problems, and that they are beginning to solve them in similar ways. The second purpose is to describe and discuss the construction of classification schedules, and their use in various types of indexing system—alphabetical and systematic—and to examine the emergence of similar trends in mechanical selection.

PURPOSE

The classification of subject matter may be carried out for all sorts of *special* purposes—to arrange books on shelves, to group inventions in patents, to classify the raw materials, intermediates and products of importance to a particular manufacturer, and so on. All such special arrangements have their particular uses and their particular problems. What I am primarily concerned with here, however, is a *system that will satisfactorily reveal the manifold inter-relations among the concepts of science*. The practical application of such a system is *the analytical index to the detailed subject matter of science*.

The complexity of detailed scientific subjects creates great difficulties in indexing, and has led to the introduction of many new, competing and complicated techniques. An American librarian, Maurice Tauber, summed the matter up a few years ago: 'We look at a library situation torn by argument and counter-proposals that extend more widely and run deeper than any other library history has shown since 1876. . . . It seems as if progress in the technical process (of cataloguing and classification) has served to complicate rather than simplify . . . it has been insufficient to prevent the so-called crisis in cataloguing.' As Tauber put it, a chaotic condition has been brought about 'by the inability to make detailed, bibliographical cataloguing, unsystematic and illogical subject heading, and close classification meet satisfactorily the problems created by vast book collections. . . . Such a state of affairs has led to some confusion and pessimism among librarians. "Frustration" and "complexity" have become bywords of speakers and writers.' Despite the 'drastic and radical' changes proposed, Tauber urged that 'criticisms are too general. . . . *It is clear that some essential factors have been generally overlooked.*' (My emphasis, B.C.V.)

It is now widely recognised that the confusion and frustration have been partly due to a failure to dig down to the roots of our methods. A basic study of methodology has been 'the factor overlooked'. It is this task which I have here attempted to initiate.

ASSUMPTIONS

The complex techniques of information retrieval now so widely discussed among science librarians and documentalists have been criticised by those who consider—contrary to Tauber—that the traditional techniques of the Dewey classification and the dictionary catalogue, polished a little, will serve all purposes. Some of these criticisms seem to be based on a lack of appreciation of what it is that modern information retrieval aims to achieve. It is as well therefore if I set out the assumptions on which the ensuing argument is founded.

(1) The typical specific subject in science and technology is a complex of several concepts, e.g. 'the influence of X-rays on the sensibility of *Bacillus coli* treated with penicillin'.
(2) There is a large number of such specific subjects, so that an information index tends to become a large assembly of them.
(3) Searches both for specific subjects and for generic aspects of them are frequent, though the relative proportion of each type of search may vary in different situations. Ideally, an index should be designed to allow of both types of search.

(4) Three kinds of generic search occur.

(a) In one kind it should be possible to extract from the index the specific subject cited in (1) by means of any individual term or combination of terms it contains—e.g. by searching for 'penicillin' or for 'X-rays and *Bacillus coli*'.

(b) In the second kind of generic search the subject should be retrieved on searching for a term which *includes* any of the terms in the subject—e.g. for 'bacteria' (includes *B. coli*) or 'antibiotic' (includes penicillin).

(c) In the third kind of generic search the subject should, if desired, be retrieved on searching for terms co-ordinate with any term in it—e.g. for 'streptomycin' (coordinate under 'antibiotic' with 'penicillin').

(5) The aim of information retrieval is to design systems which will allow such searching to be carried out as efficiently, as simply and as economically as the complexity of the problem allows.

It is perfectly legitimate for the critic, who regards such assumptions and aims as unrealistic, too difficult of achievement or too costly, to reject them in favour of simpler but less versatile indexing systems. It is not legitimate for him to admit the assumptions, commend the aims, and then complain of the complexity of the solutions offered, unless he can achieve the same aims more simply.

ACKNOWLEDGEMENTS

It will be evident to the reader that the thoughts of many other people have contributed to the ideas set down here. Acknowledgement to those whose works have been quoted and discussed is made in the bibliographies at the end of each chapter, but some special debts must be recorded here. My first interest in classification I, like so many others, owe to W. C. Berwick Sayers, whose *Manual* has been a constant source of information, and who gave me many a kindly word of encouragement. The columinous writings of S. R. Ranganathan have influenced me greatly, as will be obvious, and I had fruitful talks and correspondence with him.

The meetings of the Classification Research Group have been a recurring stimulus to the clarification of thought, and the private discussions which often follow them have been equally fruitful. In particular, D. J. Foskett has helped to formulate many ideas and has provided a valuable survey and bibliography of classification in the A.S.L.I.B. *Handbook of Special Librarianship*. I am very grateful to him for giving me permission to reproduce one of his classification schedules. Two other members of the Group, B. I. Palmer and A. J. Wells, have written a useful introduction to the *Fundamentals of Library Classification*. R. A. Fairthorne, in a series of difficult but rewarding papers, letters and talks, has introduced the Group to lattice algebra and information theory. E. J. Coates has made valuable contributions to our understanding of notation for classification, and J. Mills has crystallised techniques for constructing schedules. The critical comments of J. E. L. Farradane have been a constant spur to clear thinking. Needless to say, none of these friends is to be held responsible for any views here put forward.

I wish to acknowledge the kindness of the Director of A.S.L.I.B., and of the Editorial Board of the *Journal of Documentation*, in permitting me to incorporate material first published in that journal. I am grateful also to the publishers, for suggestions which have, I hope, improved the arrangement of the book. Lastly, I must thank Mrs. June McBennett for typing and re-typing a wearisome text with accuracy and understanding.

April 1958 B. C. VICKERY

CONTENTS

I

THE NEED FOR CLASSIFICATION

Classifying, in its simplest terms, means putting together things or ideas that are alike, and keeping separate those that are different. A classification is most usually and helpfully visualised as an inverted tree (*Figure 1.1*). Entities classified are the 'leaves' or

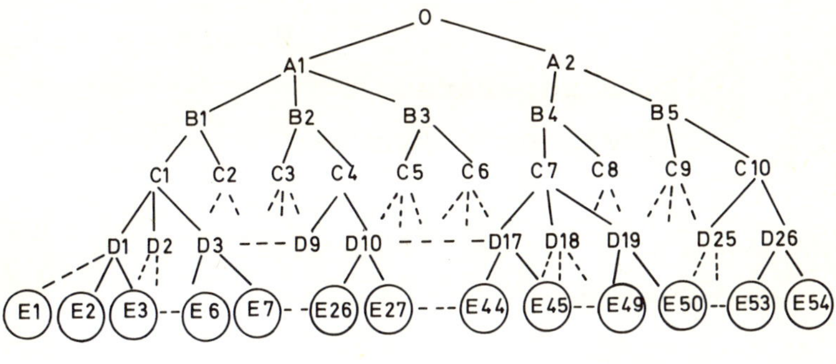

Figure 1.1 Entities (E) classified within Universe (O)

'fruit' hanging from the tree (E1, E2, etc.). Entities on a single twig are more similar to one another than they are to other entities. Groups of entities on twigs growing on the same branch are more similar to one another than they are to other groups. The most distant entities (E1 and E54) are the least alike. The nodes of the tree (at levels D, C, B, A) represent shared characteristics of entities by which they are grouped. For example, some mammals might be classified by use of the characteristics shown in *Figure 1.2*. This understanding of classification goes back to Aristotle.

The universe classified (O in *Figure 1.1*) is the highest genus, divided into species A1 and A2, which in turn are divided into sub-species, B1, B2, B3, and so on. Each of the features in *Figure 1.2* (eating habits, digits, ear auricles, hind feet, toes, horns, branching, hairiness) is a characteristic by which the universe is divided.

The basic rules of classification are:

Each characteristic of division must produce at least two classes.
Only one principle of division must be used at a time, to produce mutually exclusive classes.
The species of any genus must be completely exhaustive of their parent class.

- Eating habits
  - Carnivorous
    - Digits united
      - Ear auricles small - Sea lions
      - Ear auricles absent
        - Hind feet turn forward - Walrus
        - do not - Seal
    - Digits separate
      - Toes, only four - Cat
      - five - Bear
      - four functional - Dog
  - Herbivorous
    - Horned
      - Horns branched - Deer
      - Not branched
        - Horns hairy - Giraffe
        - Not hairy - Cattle
    - Not horned
      - Single toed - Horse
      - Eventoed
        - Four toes - Pig
        - Two toes - Camel

Figure 1.2 Characteristics used in a simple classification of some mammals

A taxonomic classification—of which *Figure 1.2* is a highly simplified example—seeks to follow these principles. However, documentation has encountered classificatory problems that cannot be solved by these principles alone.

In science there are used many classifications of entities—plants, animals, rocks, soils, stars, diseases, occupations, and so on. In these taxonomies a classification may display genetic relations—for example, an evolutionary family tree of animal species—but its prime purpose is to aid in the identification of entities. By spelling out in orderly sequence various characteristics of an animal, a rock, a disease, we make it easier to determine that it is identical with a previously described animal, rock or disease. Classification enables us to select, from a whole universe of known entities, the one that best matches one newly encountered.

CLASSIFICATION IN INFORMATION RETRIEVAL

Scientific information is faced with the following problem. On the one hand, we have the world's literature of science and technology, past and present, in many languages; on the other, an enquirer with a question. How to select, from a vast mass of words, the few that are most closely relevant to an enquiry? It is this selection process that makes use of classification.

In some forms of scientific information an explicit use of classification is easily recognised. Classified arrangement of books on library shelves is nearly universal. The reader wants to find books on similar subjects near to each other, to aid the browsing that precedes choice of a text. The shelf classification displays the structure of knowledge to the uninstructed reader—it has, we may say, a synoptic function. But at first glance it seems that many information tools designed for specific search—the location of information on particular narrow topics—make no use of classification. The indexes *within* the books on the shelves may be simple alphabetical lists of words or phrases. This impression is true enough for small indexes with a few hundred entries, but it soon ceases to be true as the size of the index increases. Even if simple word lists continue to be used, classification begins to play a role as a search aid.

Let it be clear that this is not claiming that in some way a full classification schedule lies behind all information tools (though some approximation to it may lie within the searcher's mind), but that elements of classification—often a good deal more than are immediately apparent—play an important role in all information retrieval.

A classification such as is modelled in *Figure 1.1* uses two ordering principles:

(1) By means of successive characteristics it *groups* entities.
(2) By other criteria it *sequences* them (in the sequence E1, E2, E3, adjacent entities may be more similar to each other than are separated ones).

These two principles, applied comprehensively, result in a complete hierarchical tree. But they may also be applied partially, so that a field of knowledge may be illustrated, not by a single large tree but by a collection of small shrubs. It is such a structure that is displayed by many information search tools.

There are four main areas in which classification is used in information retrieval.

The first has already been mentioned—the arrangement of books on shelves. General schemes of bibliographic classification, such as that first developed by Melvil Dewey, are most often used for this purpose. Their characteristics have been frequently described and discussed in standard texts (Sayers, Mills, Foskett, Tauber and Wise, Maltby), and some attention will be paid to them here.

The second area in which classification is directly used is subject bibliography. This can range from the classified catalogue of a library—mirroring and largely repeating the classed shelf sequence, and using the same scheme—to systematic arrangements of references to papers, reports and other documents. The classification in such subject bibliographies can be broad—sometimes no more than a sequence of small 'shrubs', as in the COSATI categories used for arranging some abstracting journals (*Table 1.1*). Many special classifications have been constructed for use of this kind, and one general scheme—the Universal Decimal Classification (UDC)—has been developed over the years with the same purpose in mind. The structures of these schemes will be considered in detail later.

Third, alphabetical indexes to subject matter (whether the matter consists of data or of references to documents in which data can be found) also make implicit use of classification, casually or systematically. 'All indexing', states John Metcalfe, 'depends directly or indirectly on alphabetical order, because whatever numbers, notations, symbols or codings it goes on to, it must begin with names written with the letters called the alphabet . . . Only known names in a known order, with direct entry under them as in alphabetical catalogues, or with references to another known order as in the classified catalogue, or to the codings of mechanical selection, will indicate information . . . Alphabetical order is the world's most important precision tool.' On the other hand, as Julia Pettee pointed out, though we must all begin our subject lists with 'the alphabetical cart before the horse, we soon discover that, though the cart must carry the load, it is the horse that must do the pulling, and the horse is the basic principle of an orderly systematic arrangement of subject materials'. Later we will examine how the alphabetical index is modified by the introduction of classification as an aid to search.

The fourth area of information in which classification is used is in what have been called 'manipulative' indexes, more often known as 'post-coordinate' systems. In these, to each item of subject matter (again, data or references) is assigned a set of terms or 'descriptors' from among those that are listed in a 'thesaurus'. At the indexing stage, at the search stage, or at both, this word list is searched to select appropriate descriptors, which are then combined in a suitable manner to represent either a subject item or a query. Choice is aided by showing relations between the terms listed in the thesaurus, and it is in the construction of these relationships that classification plays a part.

THE VALUE OF CLASSIFICATION IN RETRIEVAL

A little reflection makes evident that classification, in one form or another, at one stage or another, is almost universal in information storage and retrieval. It can be avoided at the indexing stage: an index based on document titles, for example, simply uses the 'known names' assigned to documents by authors, and lists them in a known order. Retrieval from such an index can on occasion avoid classification, if the 'name' known to the searcher happens to match a title entry. But the moment that he looks for alternative 'names' to express the topic he seeks, he has classed these names alike. As soon as we depart from the title indexing, and use a standard vocabulary of indexing terms, the indexer must of necessity class variant topic words as alike, all to be represented by a particular standard term, and the searcher must do the same. Nearly always in information retrieval, a classifying activity is implicit. But it is also often explicit.

The value of classificatory techniques in otherwise alphabetically based retrieval systems is established both by logic and by their use in practice. The alphabet is indeed a basic precision tool, but it is not the whole answer. As Peter Mark Roget wrote in 1852: '. . . words are the instruments of communication, though we often remain unconscious how much we owe to this potent auxiliary of

Table 1.1 COSATI SUBJECT CATEGORY LIST: MAIN FIELDS SHOWING SOME OF THE GROUP SUB-DIVISIONS

01 Aeronautics
- A. Aerodynamics
- B. Aeronautics
- C. Aircraft
- D. Aircraft flight control and instrumentation
- E. Air facilities

02 Agriculture
- A. Agricultural chemistry
- B. Agricultural economics
- C. Agricultural engineering
- D. Agronomy and horticulture
- E. Animal husbandry
- F. Forestry

03 Astronomy and Astrophysics

04 Atmospheric Sciences

05 Behavioural and Social Sciences

06 Biological and Medical Sciences
- A. Biochemistry
- B. Bioengineering
- C. Biology
- D. Bionics
- E. Clinical medicine
- F. Environmental biology
- G. Escape, rescue and survival
- H. Food
- I. Hygiene and sanitation
- J. Industrial (occupational) medicine
- K. Life support
- L. Medical and hospital equipment
- M. Microbiology
- N. Personnel selection and maintenance (medical)
- O. Pharmacology
- P. Physiology
- Q. Protective equipment
- R. Radiology
- S. Stress physiology
- T. Toxicology
- U. Weapon effects

07 Chemistry

08 Earth Sciences and Oceanography
- A. Biological oceanography
- B. Cartography
- C. Dynamic oceanography
- D. Geochemistry
- E. Geodesy

Table 1.1 COSATI SUBJECT CATEGORY LIST: MAIN FIELDS SHOWING SOME OF THE GROUP SUB-DIVISIONS (CONTINUED)

F. Geography
G. Geology and mineralogy
H. Hydrology and limnology
I. Mining engineering
J. Physical oceanography
K. Seismology
L. Snow, ice and permafrost
M. Soil mechanics
N. Terrestrial magnetism

09 Electronics and Electrical Engineering

10 Energy Conversion (Non-propulsive)

11 Materials
A. Adhesives and seals
B. Ceramics, refractories and glasses
C. Coatings, colorants and finishes
D. Composite materials
E. Fibres and textiles
F. Metallurgy and metallography
G. Miscellaneous materials
H. Oils, lubricants and hydraulic fluids
I. Plastics
J. Rubbers
K. Solvents, cleaners and abrasives
L. Wood and paper products
M. Corrosion and deterioration

12 Mathematical Sciences

13 Mechanical, Industrial, Civil and Marine Engineering

14 Methods and Equipment

15 Military Sciences

16 Missile Technology

17 Navigation, Communications, Detection and Countermeasures

18 Nuclear Science and Technology

19 Ordnance

20 Physics

21 Propulsion and Fuels

22 Space Technology
A. Astronautics
B. Spacecraft
C. Spacecraft—trajectories and re-entry
D. Spacecraft—launch and ground support

the reasoning faculty . . . Yet frequently we seek in vain the words we need, and the appropriate terms—notwithstanding our utmost efforts—cannot be conjured up at will. They come not when we call, and we are driven to the employment of a set of words and phrases either too general or too limited. . . .' Because of this defect—which holds for indexer and searcher as strongly as for author—Roget set about the production of his classified thesaurus of words and phrases.

He outlined its benefits in the following way: 'The inquirer can readily select, out of the ample collection spread before his eyes, those expressions which are best suited to his purpose, and which might not have occurred to him without such assistance. In order to make this selection, he scarcely ever need engage in any elaborate or critical study of the subtle distinctions (upon which the classificatory analysis has been based)—an instinctive tact will rarely fail to lead him to the proper choice . . . Some felicitous turn of expression thus introduced will frequently open to the mind a whole vista of related ideas.'

This is the very crux of the matter. An index is more than an 'indication of information'. It is a tool to aid *search*. It needs to be a 'potent auxiliary' of the information-seeking faculty. Classificatory techniques are introduced to make it something more like the 'precision instrument' that we need.

THE CHARACTERISTICS OF SUBJECT CLASSIFICATION

The most important characteristic of documentary classification is that it is concerned with *subjects*, not just with the entities of taxonomic classification. What is the nature of specific subjects—the themes on which books, parts of books, articles or parts of articles are written? A study of book titles alone would suggest that literary subjects have simple names like 'War', 'Religion', 'Boats', 'Musical pitch', 'Colour', 'Acridines', 'Wild flowers', and so on. But the study of articles at the documentation level reveals that such titles are simple in appearance only. Such a literary subject is in reality a complex aggregate of specific subjects, each of which is the main theme discussed from one particular aspect. To put the matter more precisely, the specific subject of an article is not a simple concept which can be neatly tucked away in a single pigeon-hole in the vast cabinet of knowledge, a single leaf on a hierarchical tree. It is a compound, more or less complex, of simpler concepts.

This fact is recognised in all types of system designed for information retrieval. The alphabetical index uses compound subject headings such as Electromagnetic—Waves—Propagation—Equations. Coordinate indexing systems might list each of these four words as a 'uniterm' in an alphabetical series, entering the appropriate literature references against each. In a mechanical selection system each of these terms or descriptors may be separately recorded, perhaps in coded form. An analytico-synthetic classification scheme transforms each constituent term into a classification symbol, and represents the subject by an aggregate of such symbols. All these techniques recognise the compound nature of specific subjects.

What are the elements of these compounds? Taxonomy is basically concerned with classifying 'natural kinds'—of organisms, of soils, of substances. Documentation has to classify what is written about these objects, and must take into account not only the natural kinds but also their properties, behaviour, interactions, and operations on them. Consequently, quite apart from any multiplicity of hierarchies for the natural kinds themselves, subject classification must provide hierarchies, or at least schedules, for other categories, which figure as elements in compound subjects.

For example, a classification of Soil Science, as well as hierarchies for Soil, may contain schedules for soil structure, soil constituents (chemical and biological), soil properties (physical, physico-chemical and biological), soil processes (such as formation, translocation, ion exchange, erosion), operations on soil (such as drainage, tillage, amendment), and so on.

In recent years documentalists have given a good deal of thought to bringing some order into the variety of terms that occur in hierarchies such as these. The various hierarchies that can occur in the classification of a subject field have become known as the *facets* of the subject. The technique of facet analysis is the conceptual analysis of a subject to pick out, from the theoretically unlimited number of characteristics by which a subject could be divided, those that are most significant in the study and practice of the subject. To give another example, an analysis of Astronomy might isolate the following facets, for each of which a hierarchy could be constructed:

Celestial bodies (star, comet, planet, etc.)
Their parts (axis, tail, surface, etc.)
Systems of bodies (galaxy, constellation, etc.)
Properties of bodies (size, temperature, etc.)
Properties of systems (e.g. distribution of stars)
Motion of bodies (orbit, rotation, etc.)
Relations between bodies (parallax, eclipse, etc.)
Interactions between bodies (perturbation, capture)
Operations (e.g. spectroscopic observation)
Tools (e.g. telescope)

In effect, the terms to be found within a particular subject field can be sorted into groups, each of which has been differentiated from the genus on the basis of a different characteristic. This is the sorting that is called 'facet analysis'. It consists of taking each of the terms used in a given field and defining it with respect to its parent class. For example, consider six terms from the field of Chemistry: 'alcohol', 'liquid', 'volatility', 'combustion', 'analysis' and 'burette'. The first is a *kind* of chemical substance, the second a *state* of that substance, the third a *property* and the fourth a *reaction* of it, the fifth an *operation* performed by man on it and the sixth a *device* for carrying out an operation. The six italicised terms are characteristics of division by which the terms are derived from the class Chemistry. Each of these characteristics can give rise to its own family tree of terms—alcohol, for example, is a member of the Organic compounds.

Instead of trying to construct from the original universe one vast tree of knowledge, facet analysis first groups the terms into categories—'kind', 'state', 'property', 'reaction', 'operation', 'device', and so on—and then arranges the terms within each category into the form of a classificatory tree. This is the first, and the major, difference between taxonomic and subject classification.

The ideal in taxonomic classification is to arrange all the entities of a universe into a single hierarchy: one big family tree in which each entity belongs only to a single class, each species to a single genus, and so on. Such a 'strong' hierarchy is not always possible in documentary classification.

A given object can be classed in many ways for differing purposes: a rabbit can be classed as a rodent, a fur-bearing animal, a herbivore, a pest, a pet, and so on, according to the viewpoint of the classifier. While a taxonomic classification can hope to achieve a single hierarchy, because it has a single purpose, few documentary classifications are designed to serve a single purpose: they must aid users who have varied interests. Consequently, documentary schemes must allow for cross-classification. For example, documentary classifications of Soil may comprise four or five hierarchies: soils classified by climate (arctic, sub-tropical, arid, etc.), by physiography (desert, prairie, forest, etc.), by constitution (lateritic, calcareous, alkaline, etc.), by texture (sandy, clay, loam) and by mineral origin.

It might be thought that this cross-classification is due to the failure of soil scientists to produce a generally acceptable classification, and this is partly true, but it is basically due to the varying purposes that the classification has to serve. This is shown again in the documentary classification of Chemical substances. Chemical science has long since agreed on the main basis of classifying substances—by constitution—but to serve all purposes we must also introduce classification by physical state (e.g. industrial gases), physical behaviour (e.g. gels), chemical behaviour (e.g. explosives), origin (e.g. petroleum products), mode of production (e.g. fermentation products), and even by scale of production (fine chemicals).

THE CLUSTERING OF SUBJECTS

In scientific information, then, we find that subjects—the themes and topics on which books and articles are written—cluster into fields, each of which can be analysed into its characteristic set of facets of terms. In previous pages the fields of Soil Science, Astronomy and Chemistry have been mentioned. One considerable problem is to identify appropriate fields.

Traditionally, in general classifications, subject fields have been equated with 'main classes', defined by Ranganathan as 'the fairly homogeneous, conventional regions of knowledge, which together form the first order array of classes which are mutually exclusive and totally exhaustive of the whole field of knowledge'. In turn, main classes have been identified with traditional academic disciplines—Mathematics, Physics, Chemistry, and so on. It is found that these conceptions are inadequate.

The scope of a specialised documentary classification is usually designated by its title, the subject field that it professes to classify. It is no easy task to state what is meant by a subject field. In general, it can be expressed as Thing-Activity. A definite and definable group of things (soils, celestial bodies, chemicals, food, etc.) is selected, and from the many relations in which they subsist a certain number are selected as relevant. The selected activities may be the entire range associated with the things (as perhaps in Soil Science or Astronomy), or may be more limited (as in Food Technology, where the agricultural, nutritive, hygienic and gustatory aspects of food are excluded).

When a subject field has been selected and defined, analysis will reveal its facets. It is rather rarely in documentary classification, however, that one can restrict one self to a single homogeneous subject field. The average special library, for example, serving the varied interests of its readers, will inevitably collect material on its central subject studied from marginal points of view, on tool subjects, and even on subjects quite distinct from its own which provide needed background information.

For example, in order to classify the literature of Aeronautics, it is necessary to provide a classification that includes not only central facets such as aircraft, their parts and structure, their operation and aerodynamic behaviour, but also subsidiary facets such as materials for aircraft structures, the manufacture of those materials and their physical properties, and mathematical tools for the study of aeronautics. The result is a very complex special classification.

It is easy to see that a facet that is subsidiary in one field of knowledge (as Properties of materials is in Aeronautics) is central in another subject field (in this case in Physics). This fact raises great difficulties when we come to the design of a general classification for, say, the whole of science and technology.

A general classification cannot simply be a collection of special classifications, with, for example, a slightly different assortment of physical properties appearing in each special field where some form of Material is used. A more satisfactory solution might be a single schedule of physical properties, from which terms can be drawn into combination with terms from other facets as required. Such a solution would require that we formulate a set of homogeneous facets which collectively are sufficient to express any subject in science and technology, and which avoid all redundancy; and secondly, if the product of the classification is in the form of a catalogue or a subject index, that we decide the most suitable order in which these facets should be combined to form compound subjects.

APPLICATION TO DOCUMENTATION

Classification in documentation is a tool for selection. It is essentially a 'finding system' for subject items. This function imposes on it further characteristics that are irrelevant to its functions in taxonomy.

The first of these characteristics is that it is usually necessary to apply to the scheme a shorthand notation, for easy filing and finding. A great deal of attention has been paid to the design of notations for a classified catalogue—so much so that some library classifications are known by the notational device they emphasise (the Decimal and Colon schemes). Again, the coding of schedules for use in mechanised search systems has been extensively studied by other documentalists.

A more important characteristic of a subject classification is that it is an artificial language, designed as a tool to aid in the selection of information from a store in response to search questions. The classification serves to standardise subject description, so that the descriptions of a subject used by indexer and inquirer are more likely to coincide, thus maximising the probability of finding all items relevant to an inquiry.

A theoretical classification of a subject field can list all the facets encountered in that field, all the terms in each facet, all the hierarchical relationships of each term. But such a schedule may be far too detailed in relation to the set of items to be classed. The scheme may provide a selective sieve far finer than is needed by the documentalist. Consequently, a documentary classification must reduce the theoretical detail to the minimum needed to carry out the selection demanded by searchers. Subsidiary facets that do not serve to separate one document from another can be eliminated. Specific terms that are not the subjects of independent texts can also be eliminated, or confounded with others under a more generic term. Hierarchies displaying generic relations that are not sought by inquirers can be eliminated. In short, the classificatory language must be the simplest that is needed to satisfy the questions and select from the documents available. All additional detail, no matter how theoretically or scientifically justifiable, is irrelevant.

It is worth emphasising this concept of classification as a language, because it is one that has become of increasing importance in documentation. It has lain behind the exploration into notation and coding that has been a feature of the last few decades. It is now taking on a deeper significance. We may look on a subject description as the product of a linguistic transformation. The indexer has before him a written text, and the many statements therein are transformed, by selection and condensation, into one or more brief phrases. One challenging field of research that is now emerging in documentation is linguistic analysis, whereby we may be able to formulate rules for transforming texts into descriptions. If such rules can be formulated, then the way may be open to instruct a computer to perform the task

From this point of view, a subject classification is a condensed standardised vocabulary, with a simple standardised syntax, possibly related to natural language by a set of transformation rules. It is of interest that some students of machine translation of languages consider that a standardised artificial language of this kind could serve as an intermediary in the translation from one natural language to another. This view of classification may be a far cry from that of the taxonomist, but I fancy that Aristotle, with his deep interest in both logic and language, would be vastly intrigued.

IN CONCLUSION

This introductory chapter has been concerned to illustrate the fact that classification is used in all forms of tool for information retrieval, to explain its value, to outline the structural features of subject classifications, and to indicate some of the problems that arise in their construction.

The structure of a fully developed classification can be technically described in the following way:

(1) The whole collection (universe) of entities (in this case subject terms) may be divided into a series of *fields* (for example, Physics into Mechanics, Optics, Heat, Electricity, etc.). Each field is therefore a broad grouping. The fields may be arranged in some meaningful sequence.

(2) Each field may be divided into a series of *facets* (for example, Chemistry into substances, properties, reactions, etc.). Each facet is therefore a grouping. The facets within each field may be arranged in some meaningful sequence.

(3) Each facet may be structured into a *hierarchy*, subdivided stage by stage by the application of a series of *characteristics*. These are applied in an ordered sequence. Any sequence of terms generated by successive subdivision is a *chain* (for example, the chain Machines—Mechanical and thermal machines—Nuclear reactors—Heterogeneous reactors—Thermal heterogeneous reactors—Graphite moderated—Gas cooled). Any one level of subdivision gives rise to a group of terms that constitute an *array* (for example, within the class Metals the array Bismuth, Lithium, Mercury, Potassium, Sodium). The terms in each array may be arranged in some meaningful sequence.

(4) Rules may be provided for combining (coordinating) terms from the same array, from different arrays in the same facet, from different facets in the same field, and from different fields. These rules may involve the use of special relational operators or role indicators.

(5) Each field, facet and term may be coded to fix its position in the whole system, and to facilitate unambiguous combination with other codes.

(6) An alphabetical index to the terms, facets and fields may be provided, showing the code of each.

The techniques of classificatory analysis can be used to construct a fully developed and coded classification scheme as just described.

They can also be used to structure an alphabetical word list or thesaurus—or some combination of this with classed schedules. Any of these tools may be used in either visual or machine information retrieval—at either the indexing stage or the search stage, or both. Ensuing chapters will discuss in more detail the construction and use of these tools.

BIBLIOGRAPHY

P. ATHERTON (ed.) *Classification Research.* Copenhagen, 1965

A. C. FOSKETT. *The Subject Approach to Information.* London, 2nd ed., 1971

J. C. GARDIN. 'Document analysis and linguistic theory'. *J. Docum.*, 1973, **29,** 137–68

A. GILCHRIST. *The Thesaurus in Retrieval.* London, 1971

F. W. LANCASTER. *Vocabulary Control for Information Retrieval.* Washington, 1972

A. MALTBY (ed.) *Classification in the 1970's.* London, 1972

J. METCALFE. *Information Indexing and Subject Cataloguing.* New York, 1957

J. MILLS. *A Modern Outline of Library Classification.* London, 1960

S. R. RANGANATHAN. *Prolegomena to Library Classification.* London, 3rd ed., 1963

W. C. B. SAYERS. *A Manual of Classification.* London, 4th ed., 1967, revised by A. Maltby

M. F. TAUBER and E. WISE. *Classification Systems,* Vol. 1, Part 3 of *The State of the Library Art.* New Brunswick, 1961

B. C. VICKERY. *Techniques of Information Retrieval.* London, 1970

Attention is drawn to the journal *International Classification,* which began publication by Verlag Dokumentation in 1974.

2

THE CLASSIFICATION OF A SUBJECT FIELD

The core of a special classification is a series of schedules in which the terms encountered in a given field of knowledge are grouped into distinct facets. To form compound subjects, terms from two or more facets are combined. However, the construction of schedules for a subject classification involves a good deal more than a simple sorting of terms into categories. The prior problem of designating the subject field has already been mentioned. In this chapter it is assumed that a field—a complex of things and activities—has been specified.

One approach to the specification of fields is by way of the 'things' or entities with which subjects deal. We might, for example, consider the following sample list of 'things' as defining subject fields. Each kind of thing or substance has connected with it a series of other categories—groups of terms with which it is regularly associated in compound subjects found in the literature.

'Things' Defining Subject Fields

'Ultimate' particles
Atomic nuclei
Atoms
Radicals, ions
Molecules
Molecular systems
  Homogeneous
    Solids
      Crystals
    Liquid crystals
    Surface films
    Fluids
      Liquids
      Gases
  Heterogeneous
    Mixtures
    Colloids
    Solutions
Terrestrial masses
  Minerals
  Soils
  Rocks
  etc.
Physiographic features
    Volcanoes
  Mountains
  Springs
  Seas
  etc.
Earth
Planets
  Satellites
Sun
Solar system
Stars
Star clusters
Constellation
Galaxies
Nebulae
etc.

- Cell-parts
  - Nucleus
    - Chromosomes
    - etc.
- Cells
- Tissues
- Organs
- Organ-systems
- Organisms
  - Microbes
  - Plants
  - Animals
- Communities
- Man
- Products
  - Material
    - Tools
    - Commodities
    - etc.
  - Mental
    - Logic
    - Classification
    - Mathematics
    - etc.

FACETS WITHIN EACH FIELD

The first step in constructing a special classification is to collect terms lying within a field. They may be collected from treatises, textbooks and other survey texts within the subject field; from dictionaries and glossaries; from already existing lists of subject headings; from other classifications that cover the field; from existing subject indexes. Terms can also be derived by scanning articles, reports and other literature within the field. Advice may also be obtained from subject experts in the field—for example, it is suggested by Aitchison and Gilchrist that:

(1) Experts may be asked to assist in the scanning of the literature, indexing items using the terms they consider most appropriate.
(2) Experts may be asked to list terms of importance in their subject fields.
(3) Experts may be asked to list typical questions which they might put to the system.
(4) Experts may be shown lists of terms or draft classification schemes, in their own subject fields, and asked to comment, making amendments and adding terms.
(5) Groups of experts may be asked to discuss the terminology and classification in subject areas which appear difficult to delineate and require careful clarification.

The next step is to begin to group terms into facets by assigning them to categories. Some examples of categories useful in science and technology have already been given in Chapter 1: 'substance', 'state', 'property', 'reaction', 'operation', 'device', and so on. Attempts have been and are still being made to draw up a short and ostensibly complete list of 'fundamental' categories which are applicable to any field of knowledge, or to any field of science and technology. Ranganathan, for example, recognised only five basic categories: 'personality', 'energy', 'matter', 'place' and 'time'.

Such lists of categories have been developed by the study of various fields of knowledge, and are undoubtedly helpful as a guide. But our grouping of terms in a newly approached field cannot be carried out *a priori*. New associations are continually being discovered and recorded in the literature, and we cannot too rigidly lay down in advance what *kinds* of term—i.e. what categories—will be associated. So, in order to formulate categories, our first step must be a concrete examination of the literature of the given subject. Classification must be based—as far as its raw material is concerned—on literary warrant.

In constructing a special classification which is to be used only in a particular establishment—say, the library of an industrial firm—then the most relevant literature is that which arises directly from the establishment's activities—in particular, its research and development reports. The terms and categories used in such a case will be highly selective—specific, perhaps, to the particular complex of activities which the establishment undertakes. This has led some who have had experience in constructing such classifications to feel that the only useful classifications are 'special' ones, that each special one is unique and that the very aim of a general classification is false. I do not share this view, but it does emphasise the fact that if we are to construct a generally useful classification, it must be based on a wide, non-selective survey of the literature.

To illustrate the procedure to be followed in grouping terms into categories, I will describe the study of a particular field of knowledge, Soil Science. As a representative collection of terms used in the subject, I first examined the *Multilingual Vocabulary of Soil Science* prepared by G. V. Jacks for the Food and Agriculture Organisation of the United Nations, 1954. This consists of over 350 terms arranged in 22 groups. Each term is listed and defined in eight languages. I went through the terms one by one, assigning each to a provisional category which its definition suggested. Specimen terms, their definitions, and the provisional categories to which I assigned them, are as follows:

Cohesion: the property of particles sticking together to form an aggregate. *Property*.
Gravel: particles between 20 and 2 mm in diameter. *Part* of soil.
Sticky point: maximum moisture content at which kneaded soil ceases to stick to a knife. *Measure* of soil property.
Muck: partially decomposed organic matter. *Parent Material* of soil.
Mineralisation: release of mineral matter from organic combination. *Process* occurring in soil.
Amendment: work making soil more productive. *Operation* on soil.

Proceeding in this way, I provisionally sorted the terms into 14 categories, as follows:

Category	*Example*
Soil, according to constitution	Peat soil
Soil, according to origin	Granitic soil
Soil, according to physiography	Desert soil
Part of soil	Gravel
Constituent of soil	Nitrogen
Structure of soil	Profile
Layers of soil	Horizon
Organism in soil	Bacteria
Parent Material of soil	Muck
Process in soil	Mineralisation
Property of soil	Cohesion
Measure of property	Sticky point
Operation on soil	Amendment
Equipment for operation	Plough

I next examined two existing classifications of soil science, to collect further terms and (if necessary) categories. The Colon Classification (4th edition class J:1) listed:

Soils, according to texture	Sandy clay
Soils, according to climate	Arctic soil
Substances used in amendment	Lime
Operations on these substances	Placement

The UDC expansion used by the *Bibliography of Soil Science* introduced more terms of properties, constituents, processes and organisms, but no new categories. These three sources thus gave me 18 categories altogether. Later I condensed these categories into eight facets: *kinds* of soil, *structure* (including parts and layers), *constituents* (parent material, chemicals, organisms), *properties* (including measures), *processes* and *operations* (including substances and equipment used), laboratory *techniques* and a *general* facet. The full schedules drafted for Soil Science are given in Appendix B.

A complex example of facet analysis using such special categories occurred in the field of 'Aeronautics and allied subjects.' The *things* of basic interest were Aircraft, but it was not possible to formulate a single facet for these. Aircraft can be specified by ownership, by use, by number of planes, by type of propulsion, and so on; and indeed by combinations of these (a private transport jet). Consequently, each method of specification requires a separate sub-facet to allow for their free combination (see facets 1 to 9 in the tabulation below).

The *parts* category is equally complex. There are many different types of parts in an aircraft: the body, the engine, the electrical system, and so on. Each part is similarly composite, and the sub-parts of each type are distinctive—engine parts differ from body parts. Consequently, many *parts* facets are needed. Again, aircraft and their components display many different *attributes:* the properties and behaviour of a wing differ greatly from those of a jet engine or an electrical circuit, or from those of the materials of which it is made. So, many *attributes* facets are needed. The same is true of *operations* and *agents*. Ultimately, about three dozen facets were recognised.

	Facets	*Examples*
	Aircraft:	
1.	Specified by use	Transport plane
2.	Specified by ownership	Private plane
3.	Specified by base	Land plane
4.	Specified by number of planes	Biplane
5.	Specified by type of propulsion	Jet aircraft
6.	Specified by number of engines	Twin-engine
7.	Specified by wing position	High-wing plane
8.	Specified by tail	Tailless plane
9.	Specified by take-off, etc.	VTO plane
10.	Parts of aircraft structure	Wing, Propeller
11.	Parts of parts	Propeller blade
12.	Structural parts	Spar
13.	Structural elements	Beam
14.	Components	Engine
15.	Machine components	Piston
16.	Spatial properties	Edge
17.	Behaviour of aircraft	Manoeuverability
18.	Operation of aircraft	Take-off
19.	Operational instruments	Altimeter
20.	Operating personnel	Pilot
21.	Aircraft medium	Air
22.	Behaviour of medium	Flow
23.	Properties of behaviour	Velocity
24.	Materials for aircraft specified by use	Fuel
25.	Specified by constitution	Metals
26.	Specified by form	Liquid
27.	Manufacturing operations	Extrusion
28.	Properties of materials	Elasticity
29.	Behaviour of personnel	Air sickness
30.	Behaviour of atmosphere	Humidity
31.	Engineering operations	Testing
32.	Equipment for operations	Balance
33.	Properties of equipment	Sensitivity
34.	Managerial operations	Purchasing
35.	Mathematical operations	Analysis
36.	General properties	Variation

The facets in this field can be put into six general categories:

I Things, entities—facets 1 to 9, 21
II Parts, components—facets 10 to 15

III Materials—facets 24 to 26
IV Properties, behaviour—facets 16, 17, 22, 23, 28 to 30, 33, 36
V Operations—facets 18, 27, 31, 34, 35
VI Operating agents—facets 19, 20, 32

Although the mechanical application of a set of general categories is unwise, it is undoubtedly true that it is helpful to have in mind, on approaching a new field, categories which have been used in organising the terms in other fields. The following list of general categories, developed in the course of practical facet analysis, may be helpful. (These categories are further discussed in Appendix C.)

Category	*Example*
Things, substances, entities	
Naturally occurring	Minerals, Plants, Soils
Products	Bridges, Engines, Fibres
Mental constructs	Equations, Rectangles
Their parts	Beam, Wheel, Wing
Constituents, materials	Metal, Glass, Nitrogen
Organs	Heart, Seed
System of things	
Attributes of things	
Qualities, properties, including	Cohesion, Colour, Solubility
Structure	Layers, Profile
Measures	pH, Ampere
Processes, behaviour	Vibration, Inflammation
Object of action (patient)—any thing can be a patient	
Relations between things, interactions	
Effects	Inhibition, Stimulation
Reactions	Nitration, Symbiosis
Operations on things	
Experimental	Cutting, Breeding
Mental	Calculation, Reasoning
Operating agents—any thing can act as an agent	
Properties of attributes, relations and operations	Rate, Variation
Operations on attributes, relations and operations	Measurement, Control
Place, condition	
Time	

GROUPING TERMS WITHIN EACH FACET

Within each facet we are faced with the problem of arranging the terms in a helpful order. This is the traditional problem of classification. The first point is to decide whether all the terms included in a category can be arranged in a single hierarchical tree. In the Soil Science schedule of Appendix B, facet 7, Constituents forms a single hierarchy. But in facet 9 Soils are grouped according to origin, climate, physiography, constitution or texture. Again, as shown above, Aircraft can be specified by use, ownership, base, or in half-a-dozen other ways. Each such group of terms forms a sub-facet—each forms an independent hierarchy leading down from Aircraft, or from Soil.

It would be preferable if we could choose one particular hierarchy by which to group *individual* soils, arresting the chains of classes in the other hierarchies before they reach the individuals. This is the practice in, for example, the UDC schedules for Chemistry. The main hierarchies, 546 and 547, are used to group individual chemicals, and other substance hierarchies (e.g. 541.451 Oxides, 541.48 Salts) are used only for general studies of these classes.

However, this is not possible if the field of study is still fluid. In the case of Soils, an examination of standard texts shows that there is no recognised international classification. Some well-known soil groups are based on climatic characteristics, others on physiography, others on constitution, and so on. The resulting schedule of Soils is thus a mixed cross-classification. Such a cross-classification, which is regarded by some logicians as 'worse than useless', is in fact a necessity in the systemisation of knowledge, particularly in a field where agreement on an 'essential characteristic' has not yet been reached.

This problem was discussed by Navez, in an examination of class 66 of the UDC. It appears, he wrote, that the intention has been to class chemical products on the basis of the traditional grouping of manufactures in industrial establishments. He challenged this on two grounds. First, it gives rise to multiple location: a given group of chemicals may be produced by and, hence, be classed with a whole series of manufacturing ventures. Thus benzene derivatives may be classed with dyestuffs, pharmaceutical products, plastics, perfumes, ete. Second, such groupings are continually changing. For example, traditionally soda and chlorine have been grouped together because sodium chloride served as a common raw material; but today soda can be made from sodium nitrate and chlorine is not linked with the process. In general, Navez wrote, UDC 66 reflects only the manufacturing groupings which happened to be dominant at the time of its construction. Since then, there have been sweeping changes, and the continued development of industry makes such changes incessant—so much so that the characteristic used by the UDC is quite unable to provide a schedule of long-term value.

The solution Navez put forward is to use as a primary hierarchy an arrangement based on constitution (similar to the existing UDC 546 and 547), in which should be classed all chemicals of a definite, known composition. Secondary hierarchies based on origin (e.g. Explosives), etc., would not include these individual chemicals, but would contain only general terms and such chemicals as could not be characterised by constitution.

This solution seems to be one which is often applicable to cross-classifications. Ranganathan described the primary hierarchy as being based on a 'favoured train of characteristics'. The choice of primary hierarchy, he suggested, is based on the presumption 'that the literary warrant of the favoured characteristic is greater than that of any other possible characteristics. . . . Experience shows that literary warrant favours the recognition of quite a number of non-favoured characteristics.' In order to avoid multiple location, subdivision in the secondary hierarchies 'should be stopped at the point where it lands us on a subdivision arising out of the favoured train of characteristics'. However, it is not always possible to do this.

REPRESENTATION OF ENTITIES BY COMPOUNDING CHARACTERISTICS

The sub-facets of Aircraft, listed earlier, illustrate a different way in which cross-classification may be handled. The sub-facets may be regarded, not so much as types of aircraft, but rather as *characteristics* by which aircraft may be distinguished. and which can be combined or compounded to specify a particular type of aircraft (the example of 'private transport jet' has already been cited). Another example is in the *soil* facet of Soil Science, where it is already directed that the hierarchy of kinds of soil 'by origin' should be derived by compounding with terms from the rock or mineral schedule. Similar directions could be given to form kinds by climate, physiography, constitution and texture. We could thus form the term 'Saline soil' by compounding Soil with the term 'Saline' from a Chemistry schedule. The limiting factor to such a procedure is that the Chemistry schedule would not in itself provide any means of further dividing Saline soil. The different types of Saline soil—Solonchak, Solonetz, etc.—are distinguished from each other by rather complex characteristics. Solonetz, for example, is defined as 'formerly saline soil from which the salts have been leached, with cloddy prismatic or columnar B horizon'. There are at least five attributes other than 'saline' in this definition. Is the classifier to combine all of them to represent Solonetz? If not, which is he to choose? The difficulty here is that the subject itself—soil—has not yet been scientifically systematised. Inevitably, bibliographic classification must wait until it is, and meanwhile be content to enumerate a conventional schedule of soil types.

In a field in which scientific classification has gone much further, the representation of substances by compounding their attributes is both more feasible and more useful. Recent classifications of chemical substances are moving in this direction. Instead of enumerating all individual chemical substances, some modern schemes list only the radicals or functional groups by which they are characterised. An individual substance is obtained by compounding suitable radicals. In an elementary form, this principle is used by Ranganathan for organic chemicals, thus:

	Main facet		*Second level facet*
6	Aliphatic	1	Halogen
61	Paraffin	2	Hydroxy
611	Methane	281	Ether
63	Olefine	283	Aldehyde
68	Carbohydrate	3	Acid
7	Aromatic	5	Nitrogen
711	Benzene	51	Amino
8	Heterocyclic, etc.	6	Sulphur, etc.

In this way, Methyl ether is represented as E611,281 and Amino-benzene as E711,51.

The Colon Classification has long used both synthesis and enumeration to form its *crop* facet in Agriculture. The type of crop is represented by compounding two attributes, one from each of the following schedules:

Use	*Part*
Decoration	Sap
Feed	Bulb
Food	Root
Stimulant	Stem
Oil	Leaf
Drug	Flower
Fabric	Fruit
Dye	Seed
Adhesive	Whole plant
Manure	

Thus Floriculture is represented by the combination Decorative use: Flower, and Eatable fruits by Food: Fruit. Within each group so represented, the individual crops are enumerated and symbolised in an arbitrary manner. This method could, perhaps, be used in the Soil schedule discussed earlier. But it should be noted that the attributes used to specify types of crop are not drawn from other facets. They are scheduled in the crop facet and are symbolised notationally by single digits. The compounds (such as Food: Fruit) are in fact also scheduled. This device is therefore comparable to the formation of compounds in the Colon chemical substance facet.

The general principle illustrated by these examples is that a *substance* can be represented as a compound of its *attributes*—that in choosing the semantic level of terms to be used in the classification we can go below the substance level and use only attributes. In the earlier editions of this book I suggested that it was possible that this technique should be more widely used in subject classification. This has proved to be the case.

Many instances can be found, for example, in the *Thesaurofacet* developed by Jean Aitchison and colleagues. The schedule for Nuclear Reactors lists the following characteristics:

By energy—	S4F	Thermal
	S9	Intermediate
	SA	Fast
By arrangement of fuel and moderator—	S4B	Heterogeneous
	S4C	Homogeneous
By moderator—	S4H	Graphite
	S4J	Beryllium
	S4M	Water
		etc.
By coolant—	S6	Gas
	S6H	Water
	S7	Liquid metal
		etc.

A specific reactor is represented by compounding attributes—e.g. an intermediate heterogeneous reactor with beryllium moderator, cooled by liquid metal, is coded S9/S4J/S7.

A special use for compounded attributes has been found in classifications for Engineering, where many different engineering structures may have common geometrical attributes. These may be listed as a schedule of adjectives which can be combined with an appropriate structure to qualify it. Examples of some spatial attributes are:

Portion	*Shape*	*Dimensions*	*Arrangement*
Whole	Pointed	Thin	Vertical
Part	Rounded	Thick	Horizontal
Front	Twisted	Tapered	Lateral
Back	Triangular	Low	Longitudinal
Side	Square	High	Enclosed
Top	Circular	Short	Open
Bottom	Convex	Long	Hollow
Tip	Annular		In contact
Root	Notched		Concentric
Edge	Recessed		Eccentric
Face	Curved		Asymmetric

Structures such as Wing, Plate or Ring can be combined with attributes to give such terms as 'Wing face', 'Square plate', 'Tapered wing' or 'Concentric ring'.

ORDER WITHIN EACH HIERARCHY

Any given hierarchy consists of a series of terms of increasing intension, derived from the *summum genus* by the application of a succession of characteristics. Traditional classification has laid down a number of canons which give useful guidance here. It is helpful, for example, to use well-defined and consistent characteristics of division. The derived classes in an array should be mutually exclusive and collectively exhaustive of their immediate universe. Each chain of classes should be suitable modulated, so that no links are omitted.

The terms in each array form a series of coordinate classes, and it is necessary to decide on a helpful order. The choice of order will be dictated by the nature of the subject matter, but certain general possibilities put forward by Richardson and Ranganathan are worth remembering. As summarised by Berwick Sayers, Richardson listed nine possible orders: (1) *Logical*—from the complex to the simplest; (2) *Geometrical*—according to the position of things in space; (3) *Chronological*—according to the position in time; (4) *Genetic*—according to likeness in origin; (5) *Historical*—a combination of many of the preceding characteristics; (6) *Evolutionary*—from simple to complex; (7) *Dynamic*—order of power; (8) *Alphabetical;* (9) *Mathematical*—order by means of a notation. Ranganathan listed the second, third, sixth and ninth of these principles (calling the last 'canonical' order) and added (10) Decreasing *extension* and (11) Increasing *concreteness*.

Some of the arrays in the Soil Science schedule in Appendix B illustrate these principles. For example, the groups of Biological Organisms are in evolutionary order, and the Operations facet is in roughly chronological order. Other examples are as follows:

From simple to complex
- Elements of language
  - Phoneme (vowel, consonant)
  - Syllable
  - Word
  - Phrase
  - Clause
  - Sentence
  - Paragraph
  - Composition

Spatial
  Planets (relative to Sun)
    Mercury
    Venus
    Mars
    Asteroids
    Jupiter
    Saturn
    Uranus
    Neptune
    Trans-Neptunian
Chronological
  Food technology, preliminary operations
    Transport
    Delivery, unloading
    Sorting, grading
    Trimming, peeling
    Seeding, stoning, coring
    Cleaning, washing
    Blanching
    Cutting, slicing
    Filling into containers
    Closing containers

Such principles must not be applied mechanically, if unsupported by literary warrant.

Foskett has pointed out that 'it often happens that, while the items in a facet all bear the same relation to the main subject field, they do not bear any relation to one another that would enable us to say that one item should precede or follow another'. In Food Technology 'the Products facet contains Dairy products, Sugar, Cereals, Bakery products, Edible oils and fats, Fruit and Vegetables, Meats, Fish. Apart from Cereals and Bakery products, which are related and should be kept together, there is no significant criterion by which one sequence might be judged superior to another for these groups of products.' Often, indeed, such 'disorderly' arrays may be arranged alphabetically. 'Wherever there is no preferred sequence, alphabetical arrangement can be adopted without qualms—at least until a better arrangement suggests itself.'

Every array of terms in a classification is liable to require expansion: new entities coordinate with the old may have to be added. These new entities may be terms which do not appear anywhere else in the schedules, and if there is no readily apparent position in the array to which they should be assigned, arranging them alphabetically may be the simplest course. Alternatively, if they are newly discovered or invented entities, they may be arranged chronologically.

Another form of expansion to which many arrays are liable is to be augmented by the addition of terms which already appear in other schedules of the classification. For example, in the *constituents* facet of Soil Science, new chemicals may be discovered in the soil, or new organisms; in the *property* facet, new properties may become of importance.

A general method of allowing for this occurrence is by the use of 'divide like', which Ranganathan called 'subject device'. Terms from another schedule are drawn in as required by a suitable notational device. The effect is to add them to the existing array in the order in which they appear in the other schedule. This order—as will be discussed further below—may not be the most helpful one for the expanded array, but the techniques serve as a temporary measure to allow the array to be expanded until such time as its reconstruction may be possible.

Using another classification scheme in this way needs a suitable *intercalation* device in the notation, as will be discussed later, but it may also have disadvantages. The terms in the extraneous schedule may not be listed in a helpful order—in particular, frequently occurring marginal terms may be buried in a series of rarely occurring ones. The solution in this case is to list—in the appropriate facet of our special classification—the most important *favoured* terms, and to direct that other terms be drawn in from the extraneous schedule as required.

Example

- Soil constituents
  - Chemical
    - Inorganic
      - Mineral reserve
      - Nutrient
      - Trace elements
      - Silica/sesquioxide ratio
      - Nitrogen
        - Ammonia
        - Nitrate
        - Nitrate
      - Phosphorus
      - Other (subdivide by chemical substance schedule in UDC, Colon or Bliss)

- Soil properties
  - Physical
    - Mechanical composition
    - Degree of aggregation
    - Cohesion
    - Consistence
    - Plasticity
    - Permeability
    - Porosity
      - Capillary
      - Non-capillary
    - Other (subdivide by physical properties in UDC, Colon or Bliss)

If the number of such favoured terms is small, and other terms are very rarely needed, the favoured schedule can simply be used as it stands, with no reference to the fuller schedule of another classification, provision being made to add new terms at the appropriate position alphabetically.

COMBINATION ORDER OF CATEGORIES

By establishing a definite combination order of categories in compound subjects, the position of any compound entry in the index is unequivocally determined, and the scattering of references to that compound subject is avoided. The combination order can be settled arbitrarily, but this is not the most helpful solution, because combination order serves a further purpose. It brings into juxtaposition terms which are 'interlocked' and which cannot be separated without altering the meaning of the compound subject. To establish a general combination order of categories requires that we establish persistent interlocking relations among the chosen categories.

It is not easy to establish a 'most helpful' order. With 2 categories, there are 2 possible orders; with 3 categories, 6; with 4 categories, 24; with 5 categories, 120; and so on. The difficulty of choosing the most helpful of the possible orders led Ranganathan to his concept of 'fundamental' categories of which all particular categories were manifestations. The five 'fundamental categories' he cited in an order which he claimed to be of 'decreasing concreteness', exemplified in the following series:

Category:	Personality	—	Matter	—	Energy	—	Space	—	Time
Manifestation:	Mining	—	Coal	—	Drainage	—	Wales	—	1956

Two difficulties are presented by this view. First, it is far from clear in what sense Drainage is less concrete than Mining, or even than Coal. Coal, a thing, may be more 'static' than Drainage, an operation, but is it thereby more 'concrete'? Second, of course, the categories discussed in this chapter are more numerous than those of Ranganathan, and the order of their combination needs fresh consideration.

It has been noted by Mills that the criterion of 'concreteness' is not always clear, and he introduces the idea of 'importance to the user'. For example, he writes, 'in any technology, the end product (which is the very *raison d'être* of the subject) takes precedence over the processes and operations contributing to it. The growing of a particular crop is the object of farming, so that *crop* is the chief characteristic (primary category) in Agriculture.'

This criterion of 'utility' is far more helpful than that of 'concreteness'. It has been stressed by Broadfield that every subject 'centres in ideas of the purposes it seeks to advance, and in the concepts involved in those aspects of our knowledge which are germane to particular purposes'. This acute analysis of principles is not followed by equally acute proposals for their application. Nevertheless we can agree with Broadfield that 'the selection of the purpose of a science as the character in respect of which it is classified seems theoretically and practically to offer possibilities of a satisfactory arrangement'.

How does this apply to the combination order of categories? If we have an index heading of two terms cited in the order AB, then the primary term A is (1) the term which leads into the index, and (2) the term under which all the compound headings of the type AB are collected. If on analysis we find that A is the object, the action, the end towards which B is directed, we can conclude that it is the central idea of the subject and as such should be the primary term in the heading. The analysis of purpose or utility will thus decide combination order.

The categories recognised in my analysis of Soil Science were as follows: 'kind of soil', 'part', 'constituent', 'property', 'measure', 'process', 'operation' and 'material or equipment used in operation'. The parts, constituents and properties of a given substance are clearly dependent on that substance, which forms the 'central idea' to which the other categories contribute. The parts of the soil, *qua* parts, are only examined because of the light they throw on the soil as a whole. Properties such as soil cohesion, and processes such as soil exhaustion, are investigated for the same reason. Operations such as ploughing are directed towards affecting the state of the soil, and the purpose of a tool such as a plough is to carry out the operation for which it is used. Again, the measure of a property is directed towards the property. Application of the criterion of purpose, therefore, leads us to set up chains of categories, in which successive categories are interlocked—for example:

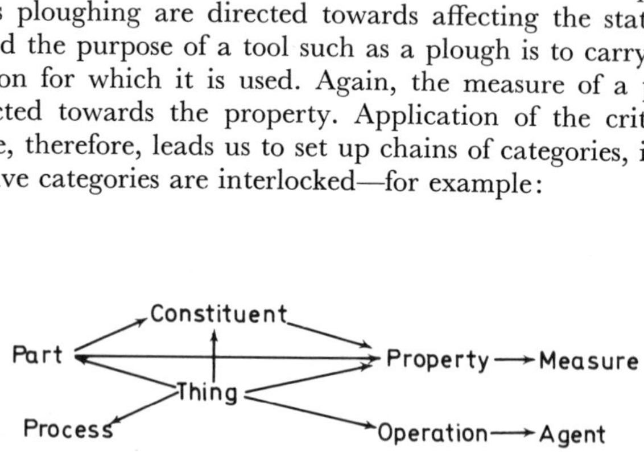

A thing has parts, constituents and properties; it undergoes processes and operations. The parts, too, have constituents and properties, and the constituents have properties, so we are led to one citation chain: (1) thing—part—constituent—property—measure. There are two other independent chains: (2) thing—process; and (3) thing—operation—agent. A decision as to the order in which these three chains are to be cited must be taken, and a criterion to aid this is required. In the schedule for Soil Science I have cited them in the numerical order given above, without any definite justification.

As an example of the type of problem encountered in establishing general rules for combination order, consider the various categories connected with the *interaction* of two entities. In such a phenomenon the following categories occur: the *reactants* (R), the *action* itself (A) and the *products* (P). When we say 'light acts on a dye to make it fade', we are distinguishing between the reactants, calling the light an *agent* (R′) and the dye a *patient* (R″). It is clear that product, agent and patient are not wholly distinct classificatory categories, since a substance such as a dye can appear now as one, now as another. The three terms express different relations between the categories of *substance* and *action*. What we have to discover is whether these relations are constant—whether the combination order of the four terms is constant, no matter what action is taking place.

In these examples of interactions between two species of substance, 'the formation of salt from soda and acid', 'the preparation of a colloidal solution of silver in water', the actions are man-made operations directed towards an end—the product. I think it will be agreed that the product should be the leading term, the combination order being P : A. If we distinguish between patient and agent, as in the subject 'the formation of salt by the action of acid (R′) on soda (R″)', then a combination order P : A : R′ seems appropriate, since the agent is directed towards the action. In language, action and agent are often united into one word, e.g. 'treatment (A) with nitric acid (R′)' is called 'nitration'.

In interactions between individuals of the same species, e.g. the fusion of two protozoa, the reactants and the products are the same. The combination order P : A can also be expressed as R″ : A. We can introduce an agent (R′); the fusion of protozoa occurs under the influence of certain conditions, so that a more general formula is R″ : A : R′. Again, in the subject 'the souring of milk', milk is very definitely not the product, it is the patient, yet equally definitely it is the focus of interest, and should be cited first: thus, R″ : A. We can again introduce an agent (e.g. bacteria), to give the combination order R″ : A : R′. Here are other examples: 'the detonation (A) of nitroglycerine (R″) by azide (R′)'; 'the corrosion (A) of iron (R″) by acid (R′)'; and 'the gravitational pull (A) of the earth (R′) on the moon (R″)'. All these conform to the combination order R″ : A : R′.

The two formulae so far developed (P : A : R′ and R″ : A : R′) can be combined in a more general combination order P : R″ : A : R′, the product (P) being omitted when it is of the same species as the patient. By combining this formula with the three citation chains already suggested, a general combination order can be offered: thing (product)—part—constituent—property—measure—patient—process/action/operation—agent, followed by place/condition—time.

I have emphasised earlier that the use of a set of 'fundamental' categories in facet analysis must not be mechanical: such a set can only serve as a guide, a reminder of a pattern that has been helpful in other situations. The same caution is necessary—indeed, much more necessary—with regard to adherence to a standard combination order. Even when agreement about the categories to be used in a given field of knowledge is obtained, there is often wide divergence among users as to the most helpful combination order.

Agreement as to interlocking relations and the occurence of chains such as thing—operation—agent is fairly readily obtained, but different users will disagree as to which link in the chain should be cited first, i.e. as to which category is from the viewpoints of 'utility' and 'purpose' the primary characteristic of their subject. Basically this disagreement is due to the fact that different users do have different purposes—e.g. a chemical engineer may want the literature of Industrial Chemistry arranged by operations rather than by substances—so that the preferred combination order must depend on the probable users of the classification. If a well-defined group of specialists is designated as the users for whom the schedules are to be constructed, a combination order suiting them can usually be decided. It is in attempting to formulate a 'most helpful' order in *general* classification that the greatest difficulty is encountered. The order suggested above can only be regarded as a general pattern which the makers of classifications may find worth considering.

TRANSFORMATION OF CATEGORIES INTO FACETS

The working out of a combination order of categories reveals the possibility of combining two or more of them in a single facet. One need for facets in classification arises as a means of regulating the order of citation of terms in a compound subject. Hence, we need to distinguish as facets in classification between two conceptual categories only if interchanging one for the other leads to a change in the helpful order of terms.

Consider a simple example. Given the four terms 'Determination' and 'Formation', 'Structure' and 'Velocity', we might call the first two *processes* A and the second two *properties* Q. Representing the category of *substance* by S, the subject (1) 'Soil:formation:velocity' gives a combination order S:A:Q. Here Velocity is a property of the process, Formation. But the subject (2) 'Soil:structure:determination' gives an order S:Q:A, since Determination is a process applied to the property, Structure. This contradiction can be resolved by introducing a fourth facet. We can call Determination an *operation* (O), so that subject (2) gives the combination order S:Q:O, facet A being vacant; or we can keep Determination as a process, and assign Velocity to another facet (G), so that Subject (1) gives the order S:A:G, facet Q being vacant. As far as this example goes, therefore, it shows that we need as facets at least four, but not more than four, of the five categories ('substance', 'process', 'property', 'determination' and the general category G).

In my schedules for Soil Science, I have in fact used all five facets. The schedules in Appendix B list both experimental manipulations (laboratory techniques) and a general facet. Let us denote these two by the symbols M and G, and the other facets as follows: kind S, structure L, constituents C, properties Q, processes A and operations O. The analysis of the previous paragraph has justified the use of the facets S, Q, A and G, but in then keeping Determination as a process it assimilates facets O and M to A. These three facets might therefore be combined into a single 'energy' facet, as is usual with Ranganathan. They occupy the same position in the combination order, and could therefore be represented by three hierarchies in a single facet.

We find, therefore, that we do not need to form facets corresponding to every conceptual category we can recognise in a subject field. Two or more categories can be 'telescoped' (as Ranganathan puts it) into a single facet if they occupy the *same* position in the preferred combination order. Again, two or more facets can be 'telescoped' into one if they occupy *successive* positions in the combination order. This is exemplified in the *operations* facet of Soil Science, which in fact comprises three successive categories: *operations* including Amendment, *substances* used in Amendment and *operations* on these substances. In such a case we can form compounds by linking successive terms from the same facet: e.g. Amendment—Lime—Placement. In the same way, in an

Engineering classification, we may combine the categories *materials* and *processes* in a single facet, forming compounds such as Steel—Heat treatment.

A further problem in the transformation of categories into facets is the combination order of sub-facets, particularly those that consist of attributes that are compounded to represent entities. Rules for deciding such order are less easy to develop than for a combination order of categories. If the primacy of 'purpose' is stressed—as discussed above—then a sub-facet such as *use* or *function* would precede a sub-facet denoting *structure:* an example would be the *crop* facet in Colon Agriculture, already discussed. 'Dependence' is another principle that may be employed. For example, it might be argued in Structural Engineering that the *method* of construction was dependent on the *design* of the structure, so that the latter sub-facet should take precedence: this sequence is used in the *Thesaurofacet*, thus:

SN	Structures
Subdivide by design	
SN6	Composite structures
SN8	Suspended structures
SNA	Surface structures
SNC	Frames
	etc.
Subdivide by method of construction	
SNP	Welded structures
SNPS	Stressed skin structures
SNQ	Monocoque structures
SNR	Honeycomb structures
	etc.

Often, however, we must rely upon literary warrant to decide how to sequence sub-facets.

DEPENDENT FACETS

It has already been mentioned that the terms in a facet may not bear any significant relation to one another except that of all being in the same category. This sometimes means that each term is followed by a chain of facets which is equally unrelated to the chains following coordinate terms. Foskett points out that in his Food Technology schedules 'the processing operations used in the cereal processing industry are, in the main, different from those in the fruit and vegetable or meat industries'. In other words, while the actual category (*operations*) is common to all the *products*, the items in it vary with the individual terms in the *products* facet.

Two solutions appear to be possible: either to make an extended operations facet listing all processes for all products in one sequence, or to make a separate facet for each product. 'The advantages of the first course are that there are usually some common processes, which need be itemised only once; further, there are some operations which, though different for each product, can be described in a sufficiently abstract way to fit several products, for example, the "removal of non-edible portions" (such as husk, chaff, peel, core, bone); and still further, it is always possible that an operation used for one product but not another may in due course be applied to the second.' For these reasons, the *operations* facet in Food Technology and also in Container Manufacture (see Appendix B) is not split up.

The opposite course is followed in Colon schedules for Agriculture. Here the second (operations) facet contains seven terms: 'Soil-ing', 'Manuring', 'Propagation', 'Disease', 'Development', 'Breeding' and 'Harvesting'. Each gives rise to a dependent chain of facets. Thus 'Soil-ing' is followed by an *operations* facet, 'Manuring' by *manures*, 'Propagation' by *organs*, and so on. Further differentiation follows: e.g. the *operations* facet after 'Soil-ing' contains four terms, each of which has its own chain of facets. The use of such a complex set of facets is justified when the terms in a facet do give rise to widely disparate dependent chains of terms.

The many facets named in the Aeronautics classification led to considerable difficulty in deciding on facet order. The basic sequence chosen is that of the categories listed in the example given before: Entities (aircraft), their parts and components, their material constituents, their properties and behaviour, operations on aircraft and on materials, and operating agents. However, the general arrangement is considerably modified because of the complex relations between facets.

At various points in the schedule the basic sequence is interrupted to insert a dependent facet which applies only to a restricted group of terms in the facet being enumerated. For example: (a) In facet 10, Parts of aircraft structure, the group 'Rotors and propellers' is followed by a new facet listing parts of rotors and propellers. (b) Similarly, the term 'Fuselage' in facet 10 is followed by a *parts* facet. (c) In facet 28, Properties of materials, the group 'Mechanical vibrations' is followed by a facet listing properties and processes relevant to vibrations. This policy of placing a qualifying facet immediately after the group of terms it qualifies leads to the apparently illogical insertion of *property* and *behaviour* facets at various points in the schedule. The reason is that facets 16 and 17 are considered to be applicable only to the earlier facets 1 to 15, and not to later facets such as 24; facets 22 and 23 are applicable to materials, and to earlier facets such as 12 or 13, but not to later facets such as 32. The *general properties* facet 36 is applicable to almost any other facet.

The same policy leads to the placing of *agent* or *instrument* facets immediately after the facets they qualify; for example, the operational instruments (facet 19) used for facets 17 and 18, and the engineering equipment (facet 32) used for facet 31.

There are times when the same facet can logically appear in more than one position in the schedule. An example of this occurs in a classification for Pharmaceuticals Manufacture. Eight facets were recognised:

A Products
B Starting materials
C Substances to be extracted
D Reactions
E Physico-chemical operations
F Agents
G Properties
H Scale of operation

Chemical substances could appear in facet A,B,C or F. The appearance of a chemical term in a particular facet is equivalent to casting it in a particular 'role'.

To sum up, an arrangement of facets is sought that will provide a schedule order that can in most cases be used automatically as a combination order.

BIBLIOGRAPHY

J. AITCHISON and others. *Thesaurofacet: A Thesaurus and Faceted Classification for Engineering and Related Subjects.* Whetstone, 1969

J. AITCHISON and A. GILCHRIST. *Thesaurus Construction.* London, 1972

A. BROADFIELD. *Philosophy of Classification.* London, 1946

D. J. FOSKETT. 'Construction of a faceted classification for a special subject', in *Proceedings of International Conference on Scientific Information,* Washington 1958

J. MILLS. *A Modern Outline of Library Classification.* London, 1960

L. NAVEZ. 'Classification des choses de la chimie'. *Rev. Docum.,* 1947, **14,** 86–91

S. R. RANGANATHAN. *Colon Classification.* London, 6th ed., 1960

B. V. VICKERY. *Faceted Classification Schemes.* New Brunswick, 1966

3

CLASSIFICATION FOR ARRANGEMENT

In science information, subject indications are used for two main purposes: first, to indicate the specific subject content of a document, and second, to arrange and synoptically display a sequence of documents on various subjects. The first function we may call 'indexing', and its relation to classification will be examined in Chapter 5. The present chapter will discuss the problems of classified arrangement. Typical examples are the arrangement of books on library shelves and of abstracts in a bibliographical publication. Typically also, the classifications needed for such purposes may be wider than the scope of a defined subject field, and so need structural characteristics above and beyond those discussed in the last chapter.

First, we must look in more detail at the problems of choosing and sequencing the subject fields within a wider, more general classification scheme. Second, the occurrence of facets in more than one field—that has already been considered from the viewpoint of a particular field—must be examined in the context of the scheme as a whole. Third, the formation of subjects by the combination of terms from more than one field gives rise to a concept that Ranganathan called 'phase relations'. Last, it is convenient to consider in this chapter the problems presented to classification by the continual changes that occur in the ways of structuring our knowledge.

FIELDS AND DISCIPLINES

In the previous chapter, in order to get started on the analysis of a subject field, I suggested that one way to define a field was to prescribe a group of 'things' and their associated categories. On p. 16 a rough and ready list of such 'things' was provided. The *field* that includes chemical substances, their states, properties and reactions, and operations carried out on them, is known as the *discipline* of

Chemistry. On the other hand, chemical substances may also appear in many other fields—e.g. as constituents in Soil Science. Here they occur only as a facet secondary to the main group of 'things', Soils, but one can find examples in which the same 'things' are a primary facet in more than one discipline—e.g. Animals in both Zoology and Agriculture. Are 'field' and 'discipline' different names for the same concept?

Mills notes that all the main current schemes of general classification distribute subjects dealing with a particular 'thing' into various classes of the scheme, according to the aspect or context in which the thing appears. For example:

Subject	*Containing class*
Sugar-beet farming	Agriculture
Sugar properties	Chemistry
Refining of sugar	Chemical technology
Nutritional value of sugar	Dietetics
Sugar for cooking	Cookery
The sugar industry	Economics

Mills regards such containing classes as fields of occupational specialisation—those of the farmer, chemist, technologist, dietician, cook and economist. The general schemes assume that this characteristic of the subjects is their most significant feature from the point of view of classification, and that the relation due to their all being concerned with the 'thing' Sugar is a secondary one.

Such fields of specialisation, Mills points out, 'are only a reflection of the most fundamental of all social institutions—the division of labour. . . . This division has a profound effect on the whole system of teaching and studying, and this is seen most clearly in the organisation of professional education. This in turn profoundly influences the literature of the subjects. As a subject is studied and taught and practised and developed, so will literature on it be produced, reflecting these activities.' And so, also, classification in its turn is based on the same specialisation. The 'disciplines' into which a general scheme is divided reflect what Bliss called 'the scientific and educational consensus'. Some historical aspects of this are discussed in Appendix A of this book.

What is clear is that the list of 'things' on p. 16 can form only a starting point in defining subject fields: at the very least, we must consider 'things in context'—i.e. the other categories regularly associated with them. Some comments on the work of the Classification Research Group on levels of entity are given in Appendix C.

Even if we define fields in this way, we cannot readily equate them with the traditional main classes, disciplines, which form the framework of education and upon which standard textbooks are written. These disciplines cannot be neatly assigned to a particular group of substances. They are often heterogeneous; they overlap; and they are sometimes not based on the 'thing' facet.

As an example of overlapping and heterogeneity, consider Physics and Physical Chemistry. These are both well-established disciplines, taught in separate university faculties, and assigned to different positions in all classification schemes. In two randomly chosen texts there are 20 chapter headings in Physics and 15 in Physical Chemistry, and 9 of these overlap, thus:

Only in Physics	*In both*	*Only in Physical Chemistry*
Mechanics	Thermodynamics	Physical properties
Head, radiation	Kinetic theory	Thermochemistry
Gravitation	Liquids	Colloids
Electrostatics	Solutions	Equilibria
Electromagnetism	Electrolytes	Chemical kinetics
Electrodynamics	Crystals	Photochemistry
Vibrations	Atomic structure	
Acoustics	and spectra	
Optics	Nuclear physics	
Relativity	Quantum theory	
Atomic physics		

The heterogeneity and overlapping of traditional disciplines has the further effect that their content varies from time to time and from place to place. There is no ambiguity, for example, about a particular subject field such as Crystals: it is built around a particular type of substance. But there is ambiguity as to its position with respect to main classes. In some contexts Physics may claim it; in others, Physical Chemistry; in yet others—both.

This ambiguity and fluidity of main classes was clearly recognised by Bliss, in his provision of 'alternative locations'. Thus Bacteriology may be placed in Botany or Biology, and Biochemistry in Biology or Organic Chemistry; Cytology may be coordinate with Morphology and Physiology, or subordinate to either; Mineralogy may be placed in Chemistry or Geology.

It would seem that the provision of alternatives is correct. Conventional main classes, traditional disciplines, in fact constitute a series of overlapping areas of knowledge whose connection with the basic classification of fields is variable and sometimes ill-defined. Since books are written about them, it is essential that a place be found for them in bibliographic classification, though it is equally essential that the basic classification should not be distorted by being forced into a framework of traditional main classes. We are faced with the problem discussed in Appendix A, in the section on the scientific and educational consensus, of including both traditional and modern divisions of the universe of knowledge in one classification.

In recent years the Indian School of classification of which Ranganathan was the centre have tried to develop the concept of 'disciplines' into that of 'basic subjects'. The most extensive survey of their work is in a series of papers by Neelameghan. The Colon Classification started with a series of traditional disciplines as its main classes—Mathematics, Physics, Chemistry, Biology, Geology, Engineering, and so on. Some of these were next divided 'canonically' to form more specialised basic subjects—for example, Physics into Properties of Matter, Sound, Heat, Radiation, Electricity, Magnetism. Two further forms of 'fission' have been recognised. One leads to subjects expounded according to different schools of thought—for example, within Psychology there are the basic subjects Gestalt Psychology, Field Psychology, Psychoanalysis. A second form of 'fission' is into subjects of restricted range, e.g. Child Medicine and Old age Medicine, or Desert Botany and Tropical Botany.

More recently, Ranganathan introduced other modes of forming basic subjects. One he called 'fusion'—giving such subjects as Astrophysics, Biochemistry, Psycholinguistics, Educational sociology. The other mode he called 'distillation', in which the theory of a practical activity has been distilled out to form a new basic subject—such as Management Science, Research Methodology, Information Science. Criteria used to identify new basic subjects have been expressed as follows: (1) the subject calls for schedules of special terms that form facets not found in other basic subjects; (2) the subject has some specialisation in academic circles.

Quite apart from the need to recognise new fields of specialisation, Ranganathan noted that there is literary warrant for the introduction of subjects wider than the traditional main classes, and he has listed what he called 'subject bundles' such as Pure Sciences, Earth Sciences, Life Sciences, Space Sciences.

It is clear that the problem of specifying the subject fields into which the universe of science is to be divided is more complex than a simple listing of 'things' suggests. Some of the fields identified by Ranganathan are indeed limitations of the 'thing' facet—e.g. *child* medicine, physics of *stars*—but other clearly recognisable specialisations are distinguished precisely by the absence of any such limitation—e.g. *heat* phenomena apply to any physical body; *research* is a wide-ranging mode of human activity. When we come to systems of thought (Gestalt and Field Psychology, or Classical Physics and Quantum Physics) we are often dealing with the same observed phenomena interpreted from different viewpoints, each with their own terminology and even using different categories—they have been called different 'cognitive sub-universes'.

No solution to these problems is offered here. Ranganathan has tried to develop the field structure of a general classification by the fission, fusion and bundling of existing traditional main classes. Austin has considerably developed the concept of a 'thing-orientated' classification (see Appendix C) but his schedules do not make contact with classes at the field level in general schemes. The International Federation for Documentation has a working committee that is, on more empirical lines, trying to establish a list of subject fields. There is much to be done before clear criteria are established.

THE SEQUENCE OF FIELDS

If an adequate list of the current subject fields of science and technology cannot yet be provided, a discussion of their preferred sequence faces difficulties. Yet there are some general points to be made.

In so far as one may define fields in terms of 'things', one might arrange the fields in some sequence determined by the complexity of the things. The list on p. 16 is in such a sequence—starting with the 'ultimate' particles of matter, building up through atoms and molecules to molecular systems, on to terrestrial and celestial masses. A second chain (on p. 17) builds up living organisms from cell-parts to communities, and last comes man and his products. Such 'integrative levels' have been much discussed by Foskett and Austin. Quite apart from the problem already mentioned—that not all fields can be based on 'things'—the development of a satisfactory set of integrative levels also faces difficulties.

An alternative approach to the sequence of fields in a classification serving to arrange documents is to be purely pragmatic. If the purpose of arrangement is to bring near to each other documents liable to be used together, then empirical data on the joint use of fields may give guidance on their sequencing. Needless to say, reliable data of this kind are hard to come by—first, because it requires the prior definition of the fields to be used and the allocation of subjects within them; second, because a sufficiently representative sample of use is difficult to obtain; and third, because use patterns are liable to change with time.

A very crude example of this approach may be quoted here. A survey by Earle and Vickery examined a representative sample of British books and articles in science and technology, and noted the publications cited in each. A two-figure class number from the Decimal Classification was assigned to each source item and each cited item. Each citation thus established a link: e.g. an item in class 52 cited an item in class 53. *Table 3.1* summarises the links thus established. For example, there were 168 instances of source class 52 citing class 53. Equally, there were 14 instances of source class 53 citing class 52. The total linkage between classes 52 and 53 was thus $168+14=182$. *Table 3.2* summarises the total linkages between classes. Over a period of time we might expect pairs of subjects to be consulted together with frequencies proportional to these total figures.

The first figure on each line of *Table 3.2* represents self-citation, e.g. class 53 citing itself. There are strong links between classes widely separated in the Decimal Classification—e.g. between 53 Physics and 62 Engineering, 54 Chemistry and 61 Medicine, 57 Biology and 61 Medicine, 62 Engineering and 66 Chemical Technology. Such a tabulation suggests that the sequence of classes in D.C. is not optimal. This particular test has many weaknesses—the subject relations were based on citations, which (like all indicators of use) are liable to bias; the citations were recorded in 1965, and use patterns may have changed; and two-figure Decimal classes are a very imperfect representation of subject fields. Nevertheless it may serve to illustrate a pragmatic approach to classified arrangement.

FACETS BELONGING TO MORE THAN ONE FIELD

A cursory inspection of the schedule for Soil Science in Appendix B shows that many terms in the lists of constituents and properties are such as would also appear elsewhere in a general classification, and indeed at several points the user is directed to use 'subject device' and subdivide by some other appropriate schedule. Such terms are common to more than one field, and the question arises, need they be enumerated more than once in a general classification?

The occurrence of terms in more than one context in a classification is well known, and is revealed in the 'relative index' to the terms in the schedules. We may take the following example from Dewey:

Water	
application in nutrition	641.16
chemistry	546.22
conservation	631.7
economic geology	553.7
engineering materials	620.198

Table 3.1 CITATIONS BETWEEN SCIENCE/TECHNOLOGY SUBJECT

Cited subject / *Source*	50	51	52	53	54	55	56	57	58	59	60	61	62	63	65	66	69	*Totals*
50	131	18	232	163	112	13	102	338	14	242	1	763	82	30	39	172	2	2 454
51	47	90	20	15	16	7	1	7	1	—	—	254	230	18	1	1	10	718
52	—	—	1 035	14	1	—	2	—	—	—	—	—	19	2	17	16	—	1 106
53	24	—	168	3 505	21	857	—	3	1	2	—	39	148	6	2	278	—	5 054
54	9	—	79	268	1 254	39	332	439	—	60	127	330	107	11	10	409	1	3 475
55	94	—	1	2	1	1 351	126	3	—	14	—	2	13	24	—	14	30	1 675
56	—	—	—	—	1	—	41	—	—	1	—	1	—	132	—	—	—	176
57	138	—	—	11	5	285	343	2 278	19	578	7	2 903	122	76	—	230	—	6 996
58	21	—	—	1	—	6	—	357	213	3	—	—	1	73	—	25	—	700
59	114	—	—	—	—	12	140	45	—	1 011	1	50	1	39	—	103	1	1 517
60	2	—	1	1	—	7	—	—	—	—	36	—	11	—	5	33	2	98
61	229	1	2	—	292	3	56	263	1	690	25	7 805	24	—	3	147	226	9 767
62	25	—	122	712	7	11	3	2	—	2	—	54	2 279	—	13	822	—	4 052
63	107	—	—	6	—	3	—	39	—	3	—	14	—	886	—	57	—	1 115
65	3	5	—	—	1	—	—	—	—	—	9	8	12	—	237	12	—	287
66	1	—	3	—	1	1	1	—	—	—	18	5	51	2	20	2 143	1	2 247
69	67	—	—	—	—	—	—	—	—	—	—	10	48	10	41	—	73	229
Totals	1 012	114	1 663	4 699	1 712	2 595	1 147	3 774	249	2 606	224	12 238	3 148	1 309	388	1 462	346	41 686

Table 3.2 TOTAL CITATION LINKAGES BETWEEN SUBJECT CLASSES

	50	51	52	53	54	55	56	57	58	59	60	61	62	63	65	66	69
50	131	65	232	187	121	107	102	476	35	356	3	992	107	137	42	173	69
51		90	20	15	16	7	1	7	1	—	—	255	230	18	6	1	10
52			1 035	182	80	1	2	—	—	—	1	2	141	2	17	19	—
53				3 505	289	859	—	14	2	2	1	39	860	12	2	278	—
54					1 254	40	333	444	—	60	127	622	114	11	11	410	1
55						1 351	126	288	6	26	7	5	24	27	—	15	30
56							41	343	—	141	—	57	3	132	—	1	—
57								2 278	376	623	7	3 166	124	115	—	230	—
58									213	3	—	1	1	73	—	25	—
59										1 011	1	740	3	42	—	103	1
60											36	25	11	—	14	51	2
61												7 805	78	14	11	152	236
62													2 279	—	25	873	48
63														886	—	69	10
65															237	32	41
66																2 143	1
69																	73

hygiene	613.41
supply	628.1
technology	661.08
utilisation	333.91

By contrast J. D. Brown in the index to his classification gives only one entry, Water, D141—and many of the correlations shown by Dewey are entered by Brown in his 'categorical tables'. This example, therefore, illustrates the two extreme methods of dealing with the problem of multiple location: Dewey enumerates Water afresh in each context in which it appears; Brown lists it only once and links it to other appropriate terms to indicate the context. We cannot assume that either solution is perfect—it is necessary to seek criteria on which a decision may be based.

A sample survey of the letter A in the alphabetical index to a volume of the *British National Bibliography* (based on Dewey) and to the indexes of the Bibliographic Classification and the Colon Classification provided over two dozen examples of multiple location.

In some of these examples, it is clear that we are not dealing with multiple location of an individual concept. 'Alcohol' is both a chemical substance and a term for a beverage which contains that substance. The same *name* is in fact applied to two distinct entities, which have different generic contexts and are rightly placed in different schedules. The chemical substance can appear as an independent facet in Chemistry, Technology, Biochemistry and Medicine; the beverage can similarly appear in Technology, Ethics and Commerce. Other examples of the same name being used to identify different concepts are 'Aerial lines' in Physics and in Aviation; 'Altitude' in Astronomy and Aviation; 'Absorption' in Light, Chemistry and Physiology; and 'Air'—a substance in Chemistry, a habitat in Ecology and an environment in Hygiene.

The situation is different in other examples: 'Aluminium' as a substance probably has a constant context of generic relations, and may thus be drawn from the same schedule whether it is produced, used or treated in various ways. 'Anatomy' no doubt means the same whether the term is applied to arthropods, cats or the human figure, and so also perhaps for 'Abdomen', 'Abyssal habitats', 'Adaptation', 'Aerobic respiration', 'Alum', 'Amber', 'Amides', 'Anaesthesia', 'Aqueducts', 'Arthropoda', 'Asbestos' and 'Astrolabe'. In each case the generic relations are probably invariable, and it is only the aspects from which the term is viewed, the other facets to which it is linked, which vary. A considerable number of multiple locations given by Dewey and Bliss would, therefore, be unnecessary in a faceted classification.

On the other hand, the assumption that the generic relations of a term are invariable must not be made too readily, without full investigation. For example, Broadfield has urged that the classification of plants in Taxonomic Botany is not suitable in other fields: in Topographical Botany an arrangement based on local distribution will be required, Pharmacology will need yet another arrangement, Commerce a fourth, and so on. Agronomy is more interested in the comparative resistance of plants to drought, cold, insects, fungi and viruses than in the comparison of their morphological characters.

This point may be further illustrated by the example of 'Animals'. These may be considered as disease carriers, as themselves diseased or as pests; they may be drawn, illustrated or photographed; and in all these circumstances their generic context may perhaps remain that of Systematic Zoology: they can be represented by a facet from this field. The terms 'cruelty' and 'welfare' relating to animals may have another generic context, and require special consideration. There remain domestic animals. Dewey, Bliss and Ranganathan list these in groupings which have no relation to zoological systems. It might be alleged that there is no logical reason why this should be so: indeed, with the growing use of scientific zoology in agriculture, logic might seem to urge the use of the same systematic arrangement in both. One reason against this is notational: by listing only certain species from the whole world of animals, shorter notational representation is achieved. This is a valuable result, which may be justified. But a more relevant argument is that to the farmer domestic livestock present characteristics and problems so different from those which are of interest to the taxonomist that they can be treated as different entities. The hierarchical order most helpful in Agriculture is quite different from that used in Systematic Zoology.

This argument has even been applied by de Grolier to such an accepted common facet as the geographical schedules. He writes that a schedule suitable for Physical Geography does not suit Economic Geography, and is even more unsuitable for History, which studies, not simply 'places', but 'areas of civilisation', 'cultural areas', political groupings and so on. He notes that Ranganathan introduced into his geographical schedules new groupings such as Atlantic countries, British Empire, underdeveloped countries, OEEC countries, and so on. In the Colon Classification these other hierarchies in the geographical facet are only used for broad groupings—specific countries are enumerated in the conventional hierarchy—but the other hierarchies might be a much more helpful way to group individual countries in certain contexts.

It seems clear, therefore, that although many examples of multiple location in traditional classifications are unnecessary, yet there also occur instances when the class contexts, the generic relations of certain concepts are variable enough to justify them appearing in more than one hierarchical order. In this case, there are three ways of dealing with them: (1) All the hierarchical orders containing the terms can be brought together in one facet, as attempted by Ranganathan in his geographical facet. (2) Each hierarchy can be enumerated as a separate facet, and can be allocated to the field in which it is most relevant—thus two animal facets are usually constructed, one for Zoology and one for Agriculture. (3) One hierarchy can be enumerated as a separate facet as fully as possible, but only the more important 'favoured' terms in each of the others, drawing other terms from the main facet—as is done in the *constituent* and *property* facets of Soil Science. This last solution allows us to list, in Soil Science, only those physical and chemical properties which are particularly significant for our interest in soil, and to enumerate them in an order helpful for this subject. On the other hand, it gives us the possibility of drawing into the facet any other properties which may from time to time be studied.

The *property* facet of Soil Science also indicates another reason why a facet which is common to more than one field may need to be enumerated more than once. Even if it were decided that the hierarchical order of properties useful in Physics and Chemistry was also suitable for Soil Science, a second problem would remain. A property such as 'moisture content' is predicable of any mass, but its particular significance for soil is revealed by the use of special *measures* of the moisture content of soil, which are not used for any other substance: moisture equivalent, sticky point, upper plastic limit, wilting point, and so on. The alternative solutions are (1) to list such special properties in a separate facet applicable only to the special substance, as is done in the Soil Science schedule, or (2) to include in a single property facet *all* special measures applicable to special substances.

Another example of this problem is the location of the *structure* facet. This is a very general term—any thing or system of things can have a structure—and it might seem that to list it only once in a general classification would be sufficient. Yet if we consider the terms which will be subsumed under Structure, we see that they will differ for every type of substance encountered. In Soil Science, for example, structure is concerned with crumbs and texture, profiles and horizons, aggregates and particles. In Chemistry, molecular structure is concerned with nuclei and electrons, valence bonds and energy levels. The *structure* facet is specific to each field of study.

To sum up, multiple location of facets is justified if either the term content of the facet or the hierarchical order within the facet varies from one context to another. It is sometimes argued that these considerations are not sufficient justification for multiple location, since the result in a classified catalogue is to scatter references to the same subject—e.g. some references to cows will be in Zoology, some in Agriculture. This argument is invalid for three reasons: (1) The alternatives from which we have to choose are to bring together in a helpful order references (a) *within* a given subject field, or (b) from *different* subject fields. The very existence of differentiated, established subject fields implies that the first alternative should have precedence. (2) The alternative is only a real one for the primary facet in a subject, e.g. the *substance* facet. Secondary facets will anyway be distributed at various points throughout the catalogue. (3) The correlation of terms from different subject fields is not discarded—they may all be brought together in the alphabetical index to the catalogue.

I have considered here facets which occur in only one field—for example, *structure*—and those which occur in several—for example, the *properties* of substances. There are also facets which seem to be common to all fields of knowledge. The most well-known of these are the 'common sub-divisions' of literary form, language, time and place (although the misgivings of de Grolier about the universality of the last have been noted). A good deal of attention has been paid to these facets in the Colon Classification and the UDC, and by Ranganathan in a series of papers on 'depth classification'.

There are, however, other groups of terms which are of very wide application. Within a given field of knowledge, the parts and properties of, the processes in and the operations on a particular kind of substance are relatively specific to that substance—thus strata, plastic limit, weathering and irrigation are relatively specific to Soil. But these properties, processes and operations themselves have properties: the plastic limit may show *variation*, the weathering may occur at a fast or slow *rate*, the irrigation is of a certain *intensity*. These *general properties* seem to be common to all fields of knowledge. Further, while operations on substances are relatively specific, there are more *general operations* such as the *measurement* of a property, the *control* of a process or an operation.

The operations applicable to some material products cover a very wide field: tools (apparatus, machines, devices, etc.) can be *designed*, *constructed*, *tested*, and so on. Some examples of all these common terms are given in the General facet at the end of the Soil Science schedules. The *Thesaurofacet* developed by Aitchison and colleagues lists a series of 'common attributes' that can be used to subdivide other classes in the schedules.

PHASE RELATIONS

Facet analysis, and the use of categories, introduces into classification a number of relations other than the generic: between a thing and its parts, its constituents, its properties, and its processes; operations on a thing and the tools and products of operation; general properties of properties, processes and operations, and general operations on them.

The formulation of such categories, however, does not exhaust the relations which are useful in classification. For example, a very common relation is that of *influence* or *effect*. This can occur between two terms in the same category, as in Soil infiltration—Effect of evaporation (*process* affected by *process*); between two categories in the same field, as in Soil salinity—Effect of irrigation (*property* affected by *operation*); or between one field and another, as in Soil—Effect of crops (*substance* affected by *substance*). Such relations have been termed by Ranganathan 'phase relations', and the term is retained here.

Ranganathan has suggested five phase relations: (1) Biasing, in which phase B is the subject towards which the exposition of phase A is biased, e.g. Mathematics for Chemists. (2) Tool, in which phase B is the tool with the aid of which phase A is studied. (3) Aspect, in which phase B is an aspect of phase A, e.g. Soil Bacteriology. (4) Comparison. (5) Influencing. A greatly extended list has been published by Perrault.

THE RESULTING STRUCTURE

The traditional view of the structure of a classification has been diagrammed in *Figure 1.1* as an inverted tree of knowledge. Division of the universe O gives rise to the array A1 and A2, division of A1 gives rise to the array B1 to B3, and so on. If the division is logically made, all the classes in each array arise from the application of a single characteristic of division—characteristic 'a', applied to O, gives rise to A1 and A2; characteristic 'b', applied to A1, gives rise to B1 to B3, and so on. A different characteristic may be used for every division or the same characteristic may be used more than once. The problems of a traditional hierarchy are (a) choice of the characteristics of division, (b) choice of the sequence in which they are to be applied, (c) choice of the sequence of terms within each array.

As has been described in this book, it is now recognised that the structure of knowledge is more complex. Hierarchically ordered areas exist, but as component parts of a wider structure that may be represented as in *Figure 3.1.* Universe O consists of an assembly

Figure 3.1 Entities within fields (SF) and facets (F)

of subject fields, SF1 to SF7, etc. Within each subject field can occur homogeneous groups of concepts known as facets, 2F1, etc. Within each facet the concepts may be ordered into a hierarchy. Every facet in the universe may be unique, or some facets may occur in more than one subject field (e.g. 2F3 and 5F3 may be identical), and indeed some facets may occur in every field. The problems of the field/facet structure are (a) choice of subject fields, (b) their sequence within the universe, (c) choice of facets within each field and (d) their sequence.

The subject of an individual book or other document may be represented in one of the following ways:

— it covers the whole universe of knowledge (O)
— it covers one subject field (e.g. SF2)
— it covers 2 or more fields (e.g. SF2+SF7)
— it may be represented as a combination of concepts (e.g. 2F2×2F3, or 2F2×5F2, or A3×A4 or more complex combinations, sometimes using explicit phase relations)

As will be discussed in the next chapter, notation is applied to a scheme in order to mechanise arrangement. In a classification that aims to display all aspects of the structure described above the notation may attempt to serve the following functions:

(a) to identify uniquely each field, facet, individual term and explicit phase relation within the scheme

(b) to sequence appropriately the terms in each array, the facets in each field, the fields in the universe
(c) to permit of the representation of each conceptual subject, of any type, that may require to be classed
(d) to sequence these subject representations appropriately, and if required permit of alternative sequences
(e) to permit of the following modes of expanding the scheme:
— further subdivision of a class (e.g. of D3)
— insertion of a new term in its appropriate place in an array (e.g. between D2 and D3)
— insertion of a new step in division (e.g. between A3 and C3)
— insertion of a new facet in its appropriate place in a field (e.g. between 2F1 and 2F2)
— insertion of a new field in its appropriate place (e.g. between SF3 and SF4)

There are two extreme methods of attempting to serve these functions. (a) The first is to represent each field, facet and term uniquely and distinctively, and to provide rules for combining numbers so as to represent conceptual topics. The result is a fully analytico-synthetic scheme, to which the Colon scheme approximates. (b) The second extreme method is to formulate in advance all expected conceptual subjects, to list each in the schedules at the point that the conceptual structure of the classification requires, and to allocate to each a notational symbol that serves the purely ordinal function of fixing its position. The result is a fully enumerative scheme, to which the Library of Congress scheme approximates. Other schemes such as those of Dewey and UDC are compromises between these extremes.

CLASSIFIED ARRANGEMENT IN SCIENCE

The uses of classification for arrangement of scientific materials are typically for books on shelves or items in a bibliographical publication. Both general and specially devised classifications are used for these purposes. In 1969 an Aslib survey looked at 1265 science indexes and current awareness services, abstracts journals and bibliographies, and analysed their use of classification, the results being shown in *Table 3.3*. Over half the sample used no classification, relying instead on alphabetical, geographical or form arrangements. Of the 44% using a classification, 37% used one privately devised, and only 7% of the sample used a general

Table 3.3 USE OF CLASSIFICATION IN SCIENCE DOCUMENTS

	Total (100%)	*No classification**	*Own classification*	*Library classification*	*Type of library classification used* UDC	*Dewey*	*Other*
Total	1265 (100%)	703 (55.6%)	473 (37.4%)	89 (7.0%)	75	9	5
Sci./Tech. (Gen.)	436	298 (68.3%)	97 (22.3%)	41 (9.4%)	32	7	2
Maths. & Nat. Sci.	38	19 (50.0%)	15 (39.5%)	4 (10.5%)	4	—	—
Astronomy	30	21 (70.0%)	7 (23.3%)	2 (6.7%)	2	—	—
Physics	54	20 (37.0%)	32 (59.3%)	2 (3.7%)	1	1	—
Chemistry	47	26 (55.3%)	19 (40.4%)	2 (4.3%)	1	—	1
Geology (inc. Geog.)	64	35 (54.7%)	21 (32.8%)	8 (12.5%)	8	—	—
Palaeontology	2	1	1	—	—	—	—
Anthropology	21	11 (52.4%)	9 (42.9%)	1 (4.7%)	—	—	1
Biology	30	15 (50.0%)	15 (50.0%)	—	—	—	—
Botany	18	12 (66.7%)	5 (27.8%)	1 (5.5%)	1	—	—
Zoology	24	13 (54.2%)	10 (41.6%)	1 (4.2%)	1	—	—
Applied Sci. (Gen.)	45	19 (42.3%)	24 (53.3%)	2 (4.4%)	1	—	1
Medicine (inc. Vet.)	183	106 (57.9%)	75 (41.0%)	2 (1.1%)	2	—	—
Engineering & Technology	185	73 (39.5%)	92 (49.7%)	20 (10.8%)	19	1	—
Chemical Industry	88	34 (38.6%)	51 (58.0%)	3 (3.4%)	3	—	—

Notes: (1) Percentages given horizontally.
(2)* Alphabetic arrangement; alpha-classed arrangement; geographical or form arrangement; no discernible arrangement.

library classification (usually the UDC). More recently, a survey of 1095 government, institutional and industrial libraries in the United Kingdom (not all scientific or technical) showed that 33% used a privately devised classification scheme, and the remainder a general library classification (40% UDC).

Noting this diversity, the 1969 Aslib survey explored the extent to which private classifications matched general schemes. Do they show merely minor divergences, or do their structures differ markedly both among themselves and as compared with general schemes? The analysis compared special schemes used in bibliographical publications, and four general classifications (Dewey, Bliss, Colon and UDC). Just as UDC was the most widely favoured scheme in science, so also it proved to be more comparable to specially devised schemes.

The most notable feature of the comparison was the diversity of the schedules, whether one compares special schemes with general, or each group within itself. One example of this is illustrated in *Table 3.4*, showing a series of classifications of general and physical chemistry. Such differences are even more marked in developing fields such as Structure of Matter or Solid State Physics. This diversity needs further discussion.

Classification plays a part in most fields of science at certain stages of their development, and is important whenever a large number of similar entities have to be distinguished (plants, animals, minerals, stars, chemicals, etc.). After an initial period of confusion and competing classifications, there is often a phase of strenuous effort to achieve standard classification, with grouping and division based on an agreed set of 'essential' characteristics—'natural' rather than 'artificial', as the argument runs. Once such a standard has settled down, however, the practical development of a science and its associated technologies often leads to the recognition that other characteristics are equally useful for certain purposes.

This process, as it applies to the classification of chemical substances, is explored in Appendix D. During the whole period up to the eighteenth century, various characteristics were used to group and divide chemicals: physical state, physical behaviour, chemical behaviour, origin, mode of production. With the developments associated with Lavoisier, constitution became firmly accepted as the basic characteristic by which chemicals should be classified. Since that time, both industry and science have begun to reintroduce earlier characteristics and add still others for certain purposes: bond type, use, scale of production.

The upshot of this is that for different purposes it may be necessary to place a given chemical in different classes: for example, Aniline

Table 3.4 CLASSIFICATIONS OF GENERAL AND PHYSICAL CHEMISTRY

Chem. Abs.

65	General phys. chem.
66	Surface chem., colloids
67	Catalysis, reaction kinetics
68	Equilibria, solutions
69	Thermo-dynamics,-chem.
70	Crystallisation and crystal structure
71	Electric phenomena
72	Magnetic phenomena
73	Spectra, optical props.
74	Radiation and photochem.
75	Nuclear phenomena
76	Nuclear technology
77	Electrochem.

Chem. Zentr.

A	General, phys. inorganic chem.
A1	Atomic nuclei
A2	Atoms, molecules
A3	Solids, liquids
A4	Thermodynamics, equilibrium grass, liquids
A5	Electrochem.
A6	Colloid chem., interfaces
A7	Kinetics, catalysis
B	Physical organic chem.
B1	Optical, electric, magnetic props., spectra
B2	Structure determination
B3	Thermodynamics
B4	Equilibria
B5	Electrochem.
B6	Kinetics, catalysis

Dewey

541.2	Theoretical chem.
.22	Molecular structure
.24	Atomic structure
.26	Stoichiometry
.28	Quantum chem.

Dewey (*continued*)

.3	Physical chem.
.34	Solutions, colloids
.35	Photochem.
.36	Thermo-chem.,-dynamics
.37	Electro-magneto-chem.
.38	Radiochem.
.39	Kinetics, mechanism
.7	Optical activity

Ref. Zhurnal

0	General
1	Theory of molec. structure and chem. bonding
2	Experimental structure determination
3	Crystal chem. and crystallography
4	Chem. of solids
5	Gases, liquids, amorphous
6	Radiochem., isotopes
7	Thermo-dynamics, -chem., equilibria
8	Kinetics, combustion, explosions, topochem, catalysis
9	Photo-, radiation and plasma chem.
10	Solutions, acids and bases
11	Electrochem.
12	Surface chem.
13	Colloid chem.

Curr. Chem. Pap.

1	Subatomics
2	Atomic and molecular structure
3	Phases
4	Surface props.
5	Colloids and macromolecules
6	Electrochem.
7	Thermodynamics
8	Kinetics, mechanism

might be classed as a Benzene derivative, an Amine, a Dye intermediate, a Toxic substance, an Organic base, and so on. The same is occurring in other fields. We are faced, in fact, with an increasing problem of cross-classification, in which an entity may fall into

Table 3.4 CLASSIFICATIONS OF GENERAL AND PHYSICAL CHEMISTRY (CONTINUED)

Bliss

CB	Physical chem.
D	Chem. constitution
E	Constitution and props.
G	Stereochem.
J	Chem. combination
O	Isomerism, etc.
S	Adsorption
T	Polarisation
CC	Stoichiometry
C/G	Atomic, molec. weights
H/J	Thermal props.
K	Solubilities
CD	Chemical dynamics
C	Statics
D	Equilibria
I	Catalysis
L	Piezochem.
M	Valency
N	Thermodynamics
P	Thermochem.
T	Colloids
U	Surface chem.
V	Solutions
CE	Electrochem.
CF	Radio-, photo-chem.

UDC

541.1	Physical chem.
.11	Thermochem.,-dynamics
.12	Chem. mechanics
.13	Electrochem.
.14	Photochem.
.15	Radiation chem.
.17	Topochem.
.18	Colloid chem.
.2	Atomic theory
.5	Valency, affinity, bonds
.6	Structure in relation to props.
.7	Allotrophy
.8	Solutions

Colon

E:	Physical chem.
21	Chem. combination (inc. Molec. structure)
22	Solution
23	Mixture (inc. Surface, colloid)
24	Thermodynamics
25	Photochem.
26	Electrochem.
27	Magnetochem.
28	Stereochem.
296	Radiochem.

Bull. Signalétique

170 A	General and physical
I/II	General
III	Determination of atomic and molec. weights
IV	Structure
V	Equilibria
VI	Solutions
VII	Diffusion
VIII	Rheology, viscosity, adherence, cohesion
IX	Chem. kinetics
X	Catalysis
XI	Combustion
XII	Thermochem.
XIII	Activity and fugacity
XIV	Radiation chem.
XV	Electrochem.
XVI	Surfaces
XVII	Colloidal and dispersed states
B	Inorganic
I	History and gen.
II	Kinetics and reaction mechanism
III	Apparatus
IV	Preparations and props.

as many classes as there are special interest groups associated with it.

We can extend the argument. A homogeneous group of entities forms a facet—for example, chemical substances. In Chemistry substances are the centre of interest, and a classification will relate them to facets such as 'structure', 'physical properties', 'reactions', 'environmental conditions'. In Biology chemical substances are of secondary interest, as cell constituents and reagents, and a classification will relate them to facets such as 'organisms', 'organs', 'tissues', 'biological processes'. In each field the same facet (chemical substances) will fit into a different pattern.

These features have led to the proliferation, for purposes of arrangement, of many special classifications, each structuring a field from its own point of view, and thus differing from each other in the grouping and sequencing of the same entities. Even a single field—such as Nuclear Science and Engineering—structured from two viewpoints that one would imagine to be similar, may present surprising divergence as to both choice of terms for inclusion, and grouping.

A general classification can adopt one of three ways of meeting this situation: (1) It can firmly place any given entity in only one class, according to what is believed to be its 'most generally useful' characteristic. A special interest can only use such a classification as a standard by abandoning any structural relations peculiar to itself. (2) It can place each entity in as many classes as special interests may dictate. It becomes then an assembly of special classifications, held together only by its alphabetical index, and in no sense a standard. (3) It can attempt some uneasy compromise between the two extremes, failing to satisfy any special interest yet annoying the purist by inconsistency.

This last is the path chosen by most general classifications, and from this arises yet another aspect of the situation. Each chosen compromise differs from the others. This is partly because each classification maker (whether individual or committee) responds differently to the clash of special interests. It is also because interests change with time, and a later classification will emphasise characteristics ignored by an earlier. As a result, each general classification tends to become trapped in a structural framework suitable only to the era in which it was made.

The relatively outmoded character of any existing general classification is particularly evident if we compare it with the most recent developments in any field. In 1961 Unesco published a survey of current trends in scientific research, prepared by Pierre Auger. His classification of material was 'based on a functional division of scientific research rather than on a purely academic classification of sciences'. Scientific research was taken to include fundamental, applied and developmental research.

The over-all division of the survey was as follows:

I Fundamental sciences (mathematical, physical, chemical, biological)
II Earth and space sciences
III Medical sciences
IV Food and agricultural sciences
V Fuel and power research (thermochemical, hydroelectric, nuclear, solar and electrical energy)
VI Industrial research (metallurgical, chemical, textile, electromechanical, transport, telecommunications, automation, construction)

Science is seen as existing not only in the disciplines traditionally so labelled but also in many areas of technology. So division III includes the science of Nutrition; division IV includes Nitrogen fixation, Ecoclimatology, Animal genetics, Fish population dynamics; division V includes Metallography, High polymers, Acoustics of buildings. The interpenetration of science and technology is far greater than is suggested by the sharp separations of general library classifications.

Even if we restrict our examination to the 'fundamental' sciences, differences betwen Auger and the general classifications are immediately apparent. The differences are illustrated in *Table 3.5*, in which Auger's 1961 classification of 'fundamental' sciences is coded according to the 1965 edition of the Dewey scheme. For example, in the main sub-divisions of the physical sciences, the traditional disciplines of Mechanics and Acoustics are missing, and Atomic Physics, Electronics and States of Matter are much more emphasised. One of the six main sub-divisions of the biological sciences is Radiobiology, a subject that barely figures in the general classifications.

At the detailed level Auger is classing, not the whole of science, but its growth points. Yet it is fair to compare his classification with the general ones, because for current and future use they, too, must cover the growth points. For example, Auger subdivides States of Matter into six sections—Fluid mechanics, Solid state, Semiconductors, Superconductors, High pressures, Low pressures. In Dewey the terms in this array are coded at very different levels in very different hierarchies.

As a second example, consider Auger's subdivision of Inorganic Chemistry. Instead of a simple division by chemical constitution, as found in the general classifications, we get: Inorganic macromolecules, Transition metal compounds, Fluorine derivatives, Compounds for electronics, Materials for astronautics, Metallurgical chemistry, Vitreous state. Again, among the 24 sections of the Biological Sciences are subjects that are found at a much more specific level in the general schemes—such as Chromosome biochemistry, Transport through membranes, Renal physiology, Cytobiology of radiation.

Table 3.5 CLASSIFICATION OF SCIENCES (AUGER), SHOWING DEWEY CLASSIFICATION PLACINGS

Classification of fundamental sciences (Auger)	*Dewey**
Mathematics: Pure and applied	510
Mathematics	510
Algebra	512
Topology	513.83
Differential geometry	516.7
Theory of functions and functional analysis	517.5
Probability and statistics	519
Automatics	510.78
Analogue computers	510.782
Digital computing	510.783
The physical sciences	530
Theoretical physics	530.1
Asymptotic behaviour as starting point of the theory	—
The non-linear spinor theory	—
Mathematical methods and quantum field theory	530.143
Possibilities of a geometric theory of pure field	530.14
Recent research on the gravitational field	521.12
Atomic physics	539.1
Quantum mechanics of the atom	530.123
Atomic and physical constants	539.14
Elementary atomic particles	539.721
Simple atomic systems	<539.14
Complex composite atomic systems	<539.14
Externally perturbed atoms	<539.14
Internally perturbed complex atomic systems	<539.14
Collisions of atomic particles	—
Plasma	537.16
Radiation and its interaction with matter	<535/537
Nuclear physics and high energy particle physics	539.7
High energy particle physics	<539.7
Medium and low energy nuclear physics	—
Experimental technique	539.73+.77
Radiant energy	535/537
Optics	535
Image formation	—
Optical materials	<535.33
Optical receptors	—
Electronics	537.5
Vacuum tubes	<537.53†
Gas discharge tubes	<537.53
Electron microscopy	<537.56
Solid state devices	621.38152
Masers	<537.5344
Parametric amplifiers	<621.381535
Principles of circuity	537.5353
Electronic instruments	621.38133
Computers	621.38195
Telecommunications	<621.38
Radar	621.3848
Heat transfer and low temperatures	<536
Heat transfer and insulators	536.2

Table 3.5 CLASSIFICATION OF SCIENCES (AUGER), SHOWING DEWEY CLASSIFICATION PLACINGS (CONTINUED)

Classification of fundamental sciences (Auger)	*Dewey**
Very low temperature physics	536.56
Magnetism and electricity	537/538
Electrostatics	537.2
Magnetism	538
Electrodynamics	537.64
The states of matter	530.4
Fluid mechanics	532
Mechanics of conventional fluids	<532
Mechanics of non-conventional fluids	—
High pressures	—
Thermodynamic properties—The molecular theory of dense fluids	<536.71
The physics of solids at high pressures	—
Dynamics of crystal lattices	—
The physics of solids at very high pressures—Allotropic transformations	<541.7
Electronic transformations and ionisation by compression at extremely high pressures	—
Very high dynamic pressures in shock waves	—
Chemical reactions at high pressures	—
Low pressures	533.5
Production of the vacuum	533.54
Measurement of low pressures	—
Composition of gases. Leaks	544.4
Solid state physics	530.41
Experimental methods	—
Structure analysis	548.8
Mechanical properties	548.84
Optical properties	548.9
Semiconductors	537.622
Work on germanium and silicon	<546.68
Fundamental devices	621.381.52
Alloys	546.37
Superconductors	537.623
Superconductivity	<537.623
Quantum theory of the conductivity of crystals	<548.85
The chemical sciences	540
General chemistry	541
Thermochemistry and chemical thermodynamics	541.36
Electrochemistry	<541.37
Chemical kinetics and combustion	541.39
Catalysis and catalysts	541.395
Analytical chemistry	543/545
Nuclear chemistry	541.38
Chemical effects of radiation	”
Radiochemistry	”
Inorganic chemistry	546
Inorganic macromolecules	<541.2254
Compounds of the transition metals	546.62/64
Fluorine derivatives	546.731
Compounds for the electronics industry	—

Table 3.5 CLASSIFICATION OF SCIENCES (AUGER), SHOWING DEWEY CLASSIFICATION PLACINGS (CONTINUED)

Classification of fundamental sciences (Auger)	*Dewey**
Non-metallic materials for astronautical purposes	—
Metallurgical chemistry	669.92
The vitreous state	—
Organic chemistry	547
Organo-metallic chemistry	547.05
Silicon compounds	547.08
Organic fluorine compounds	<547.02
The biological sciences	570/590
Biochemistry and biophysics	574.19
Intermediary metabolism	—
Photosynthesis	581.13342
Enzyme biocatalysis (enzymology)	574.1925
Chromosome biochemistry	<574.8732
Cell biology	574.87
Ultra-microscopic structure	—
Transport through membranes	<574.875
Excitability and contractility	—
Differentiation and ageing	574.8765
General physiology	574.1 or 612
Microphysiology	574.876
Renal physiology	612.463
Synthetic physiology	—
Neurophysiology	612.8
Plant biology	580
The organisation of the plant cell	581.872
Growth and development	581.3
Genetics and taxonomy	581.15
Animal biology	590
Invertebrates	592/595
Vertebrates	596/599
Radiobiology	574.1915
The cytobiology of radiation	—
Somatic effects of whole-body irradiation on multicellular organisms	612.01448
Genetic effects of radiation	575.131
The earth and space sciences	52+55
The land mass	551.1/.2
The earth crust	551.13/.14
The internal structures	551.11/.12
The hydrosphere	<551.4
The oceans	551.46/.47
Hydrology	551.48/.49
The atmosphere. Meteorology	551.5/.6
The very high atmosphere and outer space	
Aeronomics	—
Space	523.111
Astronomy—Astrophysics	520
Stars	523.8
Galaxies	523.112

* < means 'included in'. † or 621.381512.

Particular areas of science rise and fall in importance; new topics continually emerge; the class relations of current interest alter; whole fields are restructured. It has been maintained by some writers that a science can only be properly classified when it has ceased to develop. What is certainly true is that only in a slowly developing field can current documentation continue to use an old classification. The dynamic of modern science is making outmoded the structural frameworks first laid down by Dewey (1876), UDC (1905), Bliss (1935) and Colon (1933).

LEVELS OF STANDARDISATION

The makers of a general classification typically stand in front of the universe of information, classificatory knife in hand, and decide how to carve it up—into fields, facets, chains, arrays. Most of them spend a great deal of effort discussing the higher levels of the scheme—what fields will be recognised? in what sequence will they be arranged? by what principles will these decisions be justified? An historical note on this aspect of the classification of science is given in Appendix A.

In practice, most general schemes end up by using traditional disciplines as the main fields, and this was explicitly justified by Bliss as representing the 'scientific and educational consensus . . . an order likely to persist for centuries to come'. In actual fact, although the *names* of traditional disciplines do persist for centuries, their *content* (particularly in science) changes continually and rapidly.

The inadequacy of division by traditional disciplines has been discussed by writers on Physics, which is still conventionally divided into Mechanics, Acoustics, Optics, and so on. It has been noted that this classification of physics was originally based on types of sense perception, but that this method of division has lost its meaning. A century ago these sub-fields were separate sciences, in very different stages of development, each following its own line of growth independently of the others. Today the boundaries between them have almost completely gone. These comments could be illustrated from *Table 3.6:* from all the 'sense-perception' fields of Physics, material has been drawn together into a new field, Solid State Physics.

The method adopted by makers of general classifications—successive subdivision of the universe of information, successive 'expansion' of the scheme to keep pace with the growth of the

Table 3.6 SOLID STATE PHYSICS

17.05	Structure of solids	60	Electrical properties
06	Phase transformations	61	Metals, conductors
07	Surfaces	63	Superconductivity
08	Films	64	Superconducting materials and devices
09	Adsorption	66	Semiconductors
10	Non-crystalline state	67	Semiconducting materials
15	Crystallography	68	Semiconducting devices
20	Microstructure	69	Dielectrics
25	Crystal lattice structures	70	Thermoelectric properties
30	Lattice mechanics	73	Photo-conductive,-voltaic effects
32	Acoustical properties	77	Electron and ion emission
34	Thermal properties	80	Magnetic properties
37	Diffusion	81	Paramagnetic
40	Defect properties	82	Ferromagnetic
42	Colour centres	83	Ferrimagnetic
45	Radiation effects	84	Antiferromagnetic
50	Mechanical properties	85	Magnetic resonances
55	Electron states	90	Optical properties
		95	Luminescence

libraries to which it is applied—of necessity implies an attempt to seek for permanence at the higher levels of the scheme. Bliss and many other have sought for a 'permanent' set of fields, 'the order of the sciences'. Ranganathan has rather sought uniformity in a set of facets applicable to every field. Special classified arrangements, as we have noted, display great diversity. One may suspect this to be inevitable.

Each particular bibliographic document is (or should be) designed to serve the interests of a particular user group. Increasingly, as science and technology interpenetrate, user interests diversify. The discipline-oriented aids typical of the past give way to, or are complemented by, mission-oriented aids drawing material from many scientific disciplines. The list of science/technology subject categories considered important by COSATI (*Table 1.1*) is sufficient illustration of this. Each user interest makes its own selection of subject matter, and arranges it according to its own structural pattern. There are as many arrangements of fields as there are interests.

Does this mean that attempts at standardisation must be abandoned? Before so concluding, let us consider more closely the alternative approach that has been adopted by the makers of both special classifications and thesauri designed for detailed indexing rather than arrangement. Each indexing aid, in effect, defines a field not in respect to its position in a wider universe but by enumeration of its content. In special classifications for indexing, the field embraced may perhaps be expressed in a 'thing-process' phrase, such as 'diamond technology', 'food science', 'container manufacture' and so on. If we examine a typical thesaurus, we again find subject category fields defined by the terms that they include. In both cases, the term content of the field is structured hierarchically. Higher level relations *between* fields may not be displayed.

Essentially, therefore, both types of indexing aid contain many low-level hierarchies. Each hierarchy is usually restricted to a homogeneous set of terms—things, or their parts, constituents, structures, properties, actions or processes, or operations on things, or operating agents, and so on—each homogeneous set thus being a facet. Facets are the building blocks out of which special classifications and thesauri may be assembled. Each special information aid may bring together a different set of facets, and arrange them in a different pattern, but the same facet may be common to many fields, and arrangement *within* the facet may be the same in each field. It is at this level that standardisation may be possible.

ALTERNATIVES TO STANDARDISATION

In recent years, with the likelihood of continuing diversity among special classifications, an alternative to standardisation has been put forward: the possibility of developing techniques for 'switching' from one system to another.

The question is best approached concretely, by looking at the typical structure of a bibliographic unit in a documentary system. Such a unit is usually in three parts:

(1) Citation: there is a bibliographic description, which may only be a reference to a document, but which may include an abstract or other annotation.
(2) Search keys: linked to the citation are catalogue or index terms, descriptors, class numbers, etc., that are used as keys to search for individual units. The keys may be taken from document texts or their abstracts, or they may be taken from a standard terminology (either a word list such as a thesaurus or a classification schedule).
(3) Grouping and sequencing keys: in any bibliographic document in which units are meaningfully arranged, each unit must be labelled by or at any rate assigned to a particular group.

As an example, consider the following entry in *US Government Research and Development Reports* (*Figure 3.2*). The grouping key is 6T, Toxicology. The search keys are the descriptors. The rest of the entry is the citation.

In many systems there is no explicit relation between the search keys and the grouping keys. The former are much the more specific —terms for the detailed concepts that may occur in individual papers and reports—whereas grouping keys are usually much broader. Each bibliographic unit is normally assigned to only one group, though there may be cross-references to it from other grouping keys.

Suppose that we wish to draw together, into a *new* group, units from a number of existing groups. This new group might be one already existing in another information system. The only possible way of doing this clerically, with the usual type of bibliographic unit, is to define the new group as the *sum* of certain search keys, and to draw together all units labelled with those keys. An example of this may be taken from the Chemical Society Research Unit's work with *Chemical Titles*. To form a group of bibliographic units on the topic 'Transition metals', the following search keys have to

6T. Toxicology

PROBLEMS IN AERIAL APPLICATION: DIRECTION OF MILD POISONING BY ORGANOPHOSPHORUS PESTICIDES USING AN AUTOMATED METHOD FOR CHOLINESTERASE ACTIVITY,
Civil Aeromedical Inst Oklahoma City Okla
Patsy R. Fowler, and Jess M. McKenzie. Apr 67, 15p
FAA-AM-67-5

Descriptors: (*Cholinesterase, Chemical analysis), Blood chemistry, Tissues (Biology), Toxicity, Pesticides, Insecticides, Automatic, Organic phosphorus compounds, Carbamic acids, Poisoning.

An automated method, capable of measuring cholinesterase activity in blood and tissue samples, was modified to provide increased reliability. The technique was evaluated as a means of detecting and measuring the inhibition of enzyme by organophosphorus and carbamate insecticides. As many as 200 specimens a day may be analyzed by the improved method, which provides precise estimates of cholinesterase activity in normal and poisoned samples. (Author)
AD-656 211 HC$3.00 MF$0.65

Figrue 3.2 Sample bibliographic unit

be summed (an asterisk means 'irrespective of the subsequent letters'):

Iron	Ferr*	Cobalt*
Nickel*	Niccol*	Platin*
Pallad*	Osmi*	Osma*
Ruthen*	Rhod*	Irid*
Metallocen*	Group VIII	
Transition metal*		
Transition element*		
Transition series		

This illustrates what is meant here by 'no explicit relations'. The sum must be formed *ad hoc*. There is, for example, no relation between all these terms indicated by coding. In this particular example there is no classification or thesaurus which displays all the summed terms as specific to 'Transition metals'.

From the purely classificatory point of view, the ability to form new groupings of bibliographic units is the crux of the question under discussion. Quite apart from transfer between systems that use different groupings, the matter also arises within each large system: the potential facilities for regrouping units are a major reason for getting systems into a form that can be manipulated by computer.

CONCORDANCE OF TERMS

The same subject matter may be represented in different information systems by different keys. This applies both to search keys and to grouping keys. The correspondence can be displayed, and transfer of bibliographic units facilitated, by a *concordance* between two systems. To facilitate information transfer, Gardin and colleagues have developed the concept of an *intermediate lexicon:* a standard list of defined concepts, each of which could be symbolised by a code. In the subject areas for which such a code were available, each information system could label each bibliographic unit with two sets of keys: one, the set of search keys and grouping key needed for the individual system; and the other, a set of lexicon codes corresponding to those keys. A second system, desiring to extract units from the first, would concentrate on the lexicon codes, and translate them into its own set of keys. Each system, instead of having to establish a concordance with a number of others, would do so only with the lexicon. The same idea has been put forward by Lloyd, in which the intermediate lexicon is to be the UDC.

There is no necessity for such a lexicon to be a classification—it is simply a 'switching' device. In so far as it were classified, this would only be to aid the user clearly to understand the meaning and context of each concept in it. The classification would play no part in the actual switching function. A lexicon, however, could be used to switch between two classifications, and this could lead to the regrouping of bibliographic units.

One feature of an intermediate lexicon needs discussion—the specificity of its codes. In a draft lexicon produced for documentation there are 15 concepts defined, still at a fairly general level. These might be more than adequate for switching grouping keys, but not specific enough for search keys. If a key K1 in the first system is coded by the more general lexicon concept C, this will be translated by a second system into its key K2, more general than K1. For a translation preserving the original specificity the codes of the lexicon would have to be as specific as any set of search keys with which they were to be used. When this is so, each concordance becomes in fact a *synonym dictionary*. When specific system keys are translated into more general lexicon concepts, the concordance is what may be called a *generic thesaurus*.

A concordance—whether direct between systems or with an intermediate lexicon—can only be a partial solution to the problem of regrouping bibliographic units. For each system the only classificatory structure remains that which has been built into its own grouping and search keys. The 'switch' imports no structure from another system. It is therefore necessary to examine further the problem of regrouping within a single system, before taking up again the question of transfer between systems.

REARRANGEMENT OF TERMS

It has been noted that there is often no explicit relation between the grouping and search keys used in a system. Exceptions to this statement are when both sets of keys are drawn from (a) the same classified schedule or (b) the same thesaurus.

By thesaurus is meant here either the generic kind previously mentioned or the relational kind that will be described in Chapter 5. The difference may be expressed as follows: in a relational thesaurus each entry term is linked to narrower (NT), broader (BT) and related (RT) terms; in a generic thesaurus against each entry term are only BT. In either case search keys may be derived from entry terms, and grouping keys from broader terms.

In a classification schedule covering a single field it may be that any individual term occurs only once, in a single facet. Consequently, it will figure in only one hierarchical chain, and will have only one immediately generic broader term. So, although a term may be drawn into successively wider groups, by using as grouping keys terms higher up the chain, all these groups will remain part of the same hierarchy. An example is taken from a draft classification scheme in Physics:

5	Physics of aggregate matter
56	Physics of solids
561	Electrical behaviour of solids
5612	Charge carrier behaviour
56122	Scattering
561221	Mean free path

The same would be true of a thesaurus in which there was only one BT against each entry term, but a thesaurus in principle permits of the introduction of as many BT as are desired. For example, in an agricultural thesaurus the term 'Rabbit' might have as BT 'Domesticated animal', 'Animal pest', 'Pet', 'Furbearing animal', 'Rodentia', 'Meat animal'.

Detailed faceted classifications often offer the same facility, particularly as regards the classification of entities such as substances, organisms, or devices. For example, as we have seen in Chapter 2, Fixed-wing aircraft can be subdivided—

By base (Land, Sea, Carrier)
By number of planes
By type of propulsion (Propeller, Jet)
By wing position (High, Low, Mid)
By take-off (Short, Vertical)

Each of these characteristics forms a sub-facet of the Aircraft facet. A particular aircraft can be presented by compounding terms from these sub-facets (e.g. a Land-based, High-wing, Propeller-driven, Short-take-off monoplane). Each sub-facet term is thus a broader term to the specific aircraft term, and can be shown as such in an alphabetic index to the scheme. This index would thus be a generic thesaurus.

A generic thesaurus, showing the facet and sub-facets to which each term belongs, is therefore a valuable aid to grouping in an individual system. Let us now consider a system covering more than one field. The concept of 'field', in the present context, means a range of subject matter that can be represented by a string of interrelated facets and sub-facets. For example, the field of Animal Husbandry might include the

Animals by name (e.g. cattle)
Animals by use (e.g. Domesticated)
Animals by status (e.g. Breeding stock)
Animals by natural group (e.g. Bovine)
Natural attributes and processes (e.g. Lactation)
Organ systems (e.g. Reproductive)
Parts (e.g. Udder)
Diseases (e.g. Foot-and-mouth)
Disease agent, pathogen (e.g. Bacillus)
Operations (e.g. Inoculation, Feeding, Breed selection)

Fields are not necessarily mutually exclusive. A new field may be formed by the interrelation of a number of facets, some of which may already occur in other fields. The consequence of this is that a given term may have as broader terms not only its facet concept and appropriate sub-facets but also field concepts. These, too, will be valuable for forming larger groups in information systems.

TRANSFER BETWEEN SYSTEMS

Let us assume that each of a number of co-operating information systems was equipped with its own generic thesaurus, developed independently to meet its own needs, with its own choice of search keys and broader terms. From the variety of broader terms different grouping keys could be selected, so that each system could produce a variety of differently grouped and arranged bibliographic documents. (Yet other groupings would still be possible by the *ad hoc* summing of search keys, described earlier.)

The variety of broader terms available would also help other systems, giving them much more choice in selecting from the host system those groups of bibliographic units that were appropriate to their needs. Independent generic thesauri would therefore aid transfer between systems. The transferred units would still be labelled by the keys of the host system, and these might have to be translated by a concordance.

This translation step could of course be avoided if all the co-operating systems used a common set of search keys and a single generic thesaurus. There is, in my view, rather more hope of standardisation at this level than of the adoption of a standard general classification.

Search keys, in the main, are specific concepts, terms that occur within a homogeneous facet. As already suggested earlier, there is every reason to believe that systems could agree on the assignation of terms to facets and sub-facets. There can be no argument that a Rabbit is indeed Domesticated, a Pest, a Pet, Furbearing, a Rodent and a Meat animal, and all these broader terms could be assigned to the term in a standard generic thesaurus. Newly emerging sub-facets could be added. Similarly, for each new field that claims a facet, broader terms could be added.

The difficulty over adopting a standard general classification, discussed earlier, is that each information system covers a different field (or series of fields) and so organises its appropriate facets and sub-facets differently. But a generic thesaurus is not a single classification. It would display a great variety of generic relations, from which innumerable groupings and arrangements could be devised. It would be a source for many special classifications. It would be a collection of many standard elements, from which all kinds of structures could be built.

Let us look more concretely at how it could be used. One may visualise the generic thesaurus as an alphabetic list of keys. Against each key is set out all the broader terms to which it has so far been related. In the following, each term is represented as an arbitrary two-letter code:

AA — CO, LM, TY
AB — DF, GA, KL, NA, YZ
AD — MY
BA — CO
BC — SP, UL

YT — CO, AB
YZ — FU

A key such as CO would draw several others together—AA, BA, YT, etc.—and might be used as a grouping key. Hierarchical chains occur (YT>AB>YZ>FU), allowing for grouping at various levels.

A particular system might incorporate within itself only some generic relations—for example,

AA — CO, TY
AB — DF, KL
BA — CO
YT — AB
YZ — FU

If a target system wished to extract AA, BA *and* YT in the group CO, or AB in the group YZ, then an *ad hoc* summation search would be necessary, but this would be aided by the existence of the thesaurus.

The nearest approaches to a generic thesaurus as yet available are the major relational thesauri. Such a thesaurus is in fact a combined synonym dictionary ('use' and 'use for' references) and generic thesaurus (BT and their reverse, NT), together with supplementary relations (RT). Although many thesauri are not explicitly based on facet analysis, most of their hierarchies do lie wholly within one facet or sub-facet. As it stands, such a thesaurus is not wholly usable in the generic way envisaged here, but it could form the basis for development in this direction. A fully faceted classification of science could be manipulated into the form of a generic thesaurus (attention has been drawn earlier to the *Thesaurofacet*).

This discussion of alternatives to standardisation can best be summed up by setting out the situations considered, the tools that could facilitate transfer between systems, and what these tools could achieve.

(1) Each system uses its own search and grouping keys, and does not relate these keys to one another or to those of other systems. (This is the situation in systems using search keys derived from free-language texts, and a number of standard grouping keys.)

Alternative groupings within each system can be realised only by *ad hoc* summation of search keys. For one system to extract a group suitable to itself from another system requires the same *ad hoc* approach. Search keys in the source system must be translated into those used by the target system, and this needs a *concordance* between the target system and each system it uses as a source.

(2) Situation as above, except that each system also tags each bibliographic unit with one or more codes from an *intermediate lexicon.*

Grouping problems remain as before. Each target system needs only to establish one concordance, between its own keys and the lexicon codes.

(3) Each system uses search and grouping keys drawn from its own hierarchical *classification*, not necessarily related to classifications used by other systems.

Grouping at different levels within each system can be realised, but not alternative groupings. In so far as the grouping of a source system suits that of a target, transfer of groups is facilitated. Keys must be translated between systems by direct concordance. Situation (3A) is as (2)—with the general use of an intermediate lexicon to aid translation.

(4) Each system uses search and grouping keys drawn from a *standard classification.* (This situation, with UDC as the classification, holds for a number of current bibliographic documents.)

Transfer is in principle straightforward, assuming that all systems apply the classification in the same way. Alternative groupings, if desired, must be realised by *ad hoc* summation.

(5) Each system uses search and grouping keys drawn from its own *generic thesaurus*. (This situation holds for bibliographic documents using relational thesauri or subfaceted classifications.)

The extraction of alternative groupings by summation is aided, but translation by concordance is necessary.

(6) Each system uses keys drawn from a *standard generic thesaurus.* The extraction of alternative groupings is still more facilitated, and translation is unnecessary.

BIBLIOGRAPHY

P. AUGER. *Current Trends in Scientific Research.* Paris, 1961

D. W. AUSTIN. 'Prospects for a new general classification'. *J. Lib.*, 1969, **1,** 149–69

H. E. BLISS. *Bibliographic Classification.* New York, 1952–53

E. J. COATES. 'Switching languages for indexing'. *J. Docum.*, 1970, **26,** 102–10

E. DE GROLIER. 'L'étude des problèmes de classification documentaire'. *Rev. Docum.*, 1953, **20,** 105–16

P. EARLE and B. C. VICKERY. 'Subject relations in science/technology literature'. *Aslib Proc.*, 1969, **21,** 237–43

D. J. FOSKETT. *Classification for a General Index Language.* London, 1970

N. GARDIN. 'Intermediate lexicon'. *Unesco Bull. Libs.*, 1969, **23,** 58–63

A. GILCHRIST. 'Intermediate languages for switching and control'. *Aslib Proc.*, 1972, **24,** 387–99

G. A. LLOYD. 'The UDC in its international aspects'. *Aslib Proc.*, 1969, **21,** 204–08

J. MILLS. *A Modern Outline of Library Classification.* London, 1960

A. NEELAMEGHAN. 'Basic subject in Colon'. Papers in *Lib. Sci.*, 1973, **10**(2)

J. M. PERREAULT. 'Categories and relators'. *Rev. Int. Docum.*, 1973, **32,** 136–44

S. R. RANGANATHAN. 'Depth classification', in *Abgila*, 1954, in *Ann. Lib. Sci.*, 1955 on, and in *Lib. Sci.*, 1964 on

B. C. VICKERY and others. *Classification in Science Information.* London, 1969

4

NOTATION FOR THE CLASSIFIED CATALOGUE

A subject index consists, in the first place, of an assembly of references to items of information. Each reference is labelled—by a number, by letters, by words, by what we call in general a *heading*—and the headings must be related in the mind of the searcher to the subject he seeks, and so constructed or so arranged that any particular heading can be located.

An index must set out to do more than this, because the references sought, although in the index, may not bear the heading which the searcher tries to locate. There are several reasons for this. The searcher may not name his specific subject exactly; he may name a broader, narrower or allied subject, and so will look for the wrong heading; the index may record references to studies on broader, narrower or allied subjects which have a bearing on the subject sought, but these will not be revealed by the references actually located. The index must therefore be more than an orderly arrangement of labelled references: *it must also reveal the relations existing between the headings*, i.e. the relations existing in the subject matter indexed.

The various auxiliary devices of alphabetical indexing—cross-reference, subdivision, inversion—are all concerned to reflect, in the structure of the index, the relations subsisting in the subject matter. They are necessary because the basic structural feature of the index—the alphabetical *indexing order* of the headings—is inadequate to reveal these interrelations. In classified indexes, which depart from the alphabetical order, the attempt is sometimes made to display the relations solely by means of a unique indexing order: this is true of the main enumerative book classifications. Other classified indexes, however, rely also on other devices to reveal relations not indicated by the indexing order.

The order of a classified index is maintained by the use of specially designed headings—a notation—defined by Bliss as 'a system of marks or symbols in some order, denoting terms or members of a series or system of things'. For the purposes of classification, he wrote, a notation 'is *requisite* to maintain the systematic order in schedules and on shelves, and to locate them in this order'. The notation is thus correlative, complementary and subsidiary to the classified order itself. The classified order reflects, well or ill, the real order of things; notation preserves and mechanises this classified order. 'The classification is the main thing, and the notation, however real its service, does not make the classification, though it may mar it.'

Notation has more to do than to preserve and mechanise a classified order. Its function is also to reveal that order to those who are imperfectly familiar with it. It does more than enable us to express concepts succinctly, within an economic limit of the few factors which can be inscribed on the spine of a book, or to manipulate concepts in a mechanised, routine manner. It points the way to previously unsuspected relations. Conversely, the lack of an adequate notation limits our ability to reveal significant relations in the subjects symbolised. Hence its importance in the present study. In this chapter only a condensed account of some basic features of notation is offered. For more extended studies, the papers of Coates, Cordonnier, de Grolier, Foskett and myself, cited in the Bibliography, should be consulted.

THE FUNCTIONS OF HEADINGS

If a heading is to be related in the mind of the searcher to the subject he seeks, it must of necessity *name* the subject of the reference. Any symbol which has a distinct connotation suffices for this: for example, atebrine, mepacrine hydrochloride, 2-chloro-7-methoxy-5-(α-methyl-ω-diethyl-amino-butylamino)-acridine dihydrochloride, or a certain standard formula, all serve this function equally precisely with respect to this particular chemical compound. It is essential therefore, that any indexing system should include among its headings the names of the subjects indexed, inscribed in some standard language familiar to the searcher (a natural language, a chemical formula, etc.). The problems raised by this requirement are not notational: they are concerned with the provision of an index, in some standard language, to the notational headings, and with the standardisation of nomenclature and elimination of synonyms in this index.

The first notational problem is how we are to *locate* each heading among the whole mass of others. We can of course scan a randomly arranged collection until we recognise the term we seek, and this is one of the things which high-speed mechanical selection enables us to do. Visual scanning of such a mass of symbols is, however, usually too slow, so for convenience it is necessary to arrange the terms in some conventional order. This necessity imposes certain limitations on the symbolic pattern of each term. There is, for example, no accepted method of arranging symbols such as structural chemical formulae, and it is with a view to overcoming this difficulty that Dyson and others have developed cipher notations. Dyson expresses the structural feature of the molecule by 'translating, detail for detail, the features of the ideograph or structural formula into a *linear* arrangement of symbols' (letters and numbers), so that the terms could be put into a conventional order.

For an order to be conventional it must be memorable, and it is therefore usual for it to be based, not on the meaning of the symbols, but on their structure, which may or may not be related to their meaning. Now if each symbol is a unique and independent ideograph, we need as many symbols as there are independent concepts to be arranged, and it becomes necessary for the searcher to remember a conventional order of a potentially infinite number of terms. The tax on the knowledge and memory of the searcher is therefore as great as in the arrangement according to meaning.

Suppose, however, that we choose a limited number of marks or digits by combination of which we can produce a vast number of symbols. The load on the memory is reduced enormously. This is the decisive step taken in adopting the alphabet and numerals, thus confining the number of separate digits used in constructing orders for these marks (A to Z, and 0 to 9) which are familiar to every literate member of the community. These digits form the basis of modern symbolism, though of course the conventional order need not be restricted to any one set of digits—e.g. many classifications use both letters and numerals, non-Latin alphabets may be introduced, or other marks.

Index headings, however, have a third function, that of *indicating relations* between the subjects they represent. These relations are of two kinds: within a heading and between headings. Relations within a heading can only occur if it is composite—a compound of the type described on p. 50. Relations between headings can be indicated in two ways. The first is so to construct the symbols that there is a parallelism or similarity between the pattern of the symbols and the pattern of the concepts they represent: a search for related concepts then reduces to a search for similar symbols. Thus the relation between 511 Arithmetic and 5111 Numeration is revealed by the presence of the common pattern 511.

The second method of relating headings is acually to tie them together in some way—'syndesis', as Cutter called it. There are two variants of this: to tie the symbols by juxtaposition, or with the aid of a third symbol. A familiar example of juxtaposition is the arrangement of verbal headings in a classification schedule. Under a class term sub-classes are listed, indented to the right. However, it is not simply the positioning of the later terms below the main class which tells the reader that they are sub-classes. He must also be aware of the meaning of the indentation, which is itself a *symbol* of logical division. On the other hand, let us take the entry in an alphabetical index: 'Alkali metal, *see also* Lithium, Sodium, Potassium'. The main term is linked to its subsidiaries by a third symbol, *see also*, but the mode of linkage is—juxtaposition. Therefore in more general terms we can say that this method of showing relation between two headings is to juxtapose each of them to a third symbol, the latter representing the form of relation itself.

The importance of revealing relations between headings comes out when we consider the strategy of information retrieval. The index headings do more than name subjects and list them in a preferred order: they also indicate the route to be followed in locating a specific subject. Consider the following series of headings, constructed from the Soil Science schedules given in Appendix B:

(1)	SS9i	Subtropical and tropical soils
(2)	SS9ib	Laterite soils
(3)	SS9id	Reddish brown soils
(4)	SS9id6	Physical properties
(5)	SS9id6p	Moisture content
(6)	SS9id6p3b	Measurement
(7)	SS9id6s	Salinity
(8)	SS9id8	Structure
(9)	SS9ih	Groundwater
(10)	SS9ij	Red earth

The searcher who has consulted the alphabetical index to the catalogue to discover what it contains on 'Laterite soils' will enter it at heading (2), SS9ib. He will want to know (a) which of the succeeding entries refer to laterite soils, and (b) which of the preceding may include information on them. In more general terms, he will want to know the hierarchical relations between the headings. These relations must be displayed by the headings themselves, either in the notation or in some other structural feature of the heading (e.g. indentation, styles of writing). In the above example all the hierarchical relations are indicated by indentation of the verbal headings. In headings 2, 4, 5, 6, 7 and 8 hierarchical relations are also indicated by the addition of another digit to the notational symbol (thus SS9id6 includes SS9id6p). In headings 3, 9 and 10 the notation does not indicate hierarchy (SS9ib in fact includes

SS9ih). The two types of notation are known respectively as 'hierarchical' and purely 'ordinal'.

THE MAINTENANCE OF ORDER

The following schedules, drawn from the Colon Classification, will serve to illustrate the structure of a faceted classification. The notation shown is based on that of Colon.

Main class: U, Geography
Facet I: Geographical divisions

1	World
1–52	Roman empire
1–56	British empire
11	Eastern hemisphere
12	Southern hemisphere
15	Countries bordering Indian ocean
16	Countries bordering Atlantic ocean
191	Equatorial zone
192	Tropical zone
1N	League of Nations area
1N4	United Nations area
1(P111)	English-speaking countries
1(Q7)	Moslem countries
4	Asia
41	China
4111	Kiang-su
419B	East China
419M	West China
42	Japan
49B	East Asia
4(Q7)	Moslem countries in Asia
5	Europe
56	British Isles
561	England
56114	Hertfordshire
6	Africa
7	America
8	Australia
91	Land within Indian ocean
92	Land within Atlantic ocean
95	Indian ocean
96	Atlantic ocean

Second-level facet: Physical features

1	Land (geosphere)
2	Mountain
3	City
5	Water (hydrosphere)
etc.	

Facet II: Geographical problems

1	Mathematical geography	7	
2	Physical geography	8	
21	Geomorphology	9	
25	Oceanography	92	
28	Meteorology	93	
281	Weather	94	Ordinal
285	Humidity	95	notation
2855	Rainfall	96	
287	Climate	97	
3	Biogeography	98	
4	Anthropogeography	991	
5	Political geography	992	
6	Economic geography	993	

Main class: V, History

Each main class is divided into facets. The first array of terms within a facet are all of the same category, but may be derived on the basis of various characteristics. The terms derived by each independent characteristic are in a hierarchical arrangement, which is displayed by the indentation of the verbal headings. Let us consider the probmles of indexing order which such a classification raises.

(1) *Within a given hierarchy*, it is required that all the sub-classes of a particular term should follow it immediately: thus Weather, Humidity and Climate follow Meteorology in facet II. In the hierarchical notation of Colon, this is achieved by subdividing the schedule symbol for Meteorology, 28, by further digits to give decimal fractions 281, 285, 287. These immediately follow 28 in ordinal value. Moreover, within a given array the classes must follow one another in a particular order. This is achieved by assigning the final digits (1, 5, 7 here) in the required order. These requirements can be achieved by ordinal notation, in which the symbols given to Meteorology, Weather, Humidity and Climate (say 93, 94, 95, 97) have the correct ordinal relations, though they do not reflect in their structure the hierarchy of classes.

Both notations achieve this result by using 'significant' symbols which are decimal fractions, in which, for example, 281, 285 and 287 all preceded 29 in ordinal value. The expression 'decimal fraction notation' is commonly used for all notations which adopt this principle, though the more general term, covering other bases as well as the decimal, is 'radix fraction'.

Various ordinal radix fraction notations are compared below.

(1)	(2)	(3)	
7	F	S	Mathematical geography
8	G	T	Physical geography
9	H	U	Geomorphology
92	I	UK	Oceanography

93	J	UL	Meteorology
94	K	UM	Weather
95	L	UN	Humidity
96	M	UP	Rainfall
97	N	US	Climate
98	P	UT	Biogeography
991	Q	UTE	Anthropogeography
992	R	UTI	Political geography
993	S	UTO	Economic geography

A decimal notation is shown in column (1). The adoption of a letter base, as in (2), produces shorter symbols. Column (3) shows a notation which is designed to produce symbols whose pronunciation is internationally recognised. It has been designed by de Grolier, who uses only the letters A, E, I, K, L, M, N, O, P, S, T and U. Vowels and consonants alternate, though he allows some diphthongs, AE, AI, AU, EA, OA, OE, and perhaps some double consonants, MP, NK, NL, NS, NT.

(2) *Within a given facet*, it is necessary that independent hierarchies be kept separate, and should follow each other in a particular order. With ordinal notation this is readily achieved by assigning to successive hierarchies groups of symbols in ascending order. For hierarchical notation the situation is more complex.

Facet I of the schedule on p. 77 shows various methods of solving the problem. In the schedule the world is divided geographically into ten independent hierarchies: (a) physical features, (b) empires, (c) hemispheres, (d) countries bordering oceans, (e) climatic zones, (f) supra-national groupings, (g) subject groupings, (h) national divisions, (i) countries within oceans, and (j) oceans. Ranganathan decided that these should be arranged in the order given, and has achieved this by the following devices. Hierarchy (a) is introduced by a connecting symbol, the dot, which is lower in ordinal value than any other symbol in the illustrative schedule. The remaining hierarchies are represented by symbols of successively higher ordinal value: 1–, 11 to 14, 15 to 18, 19, 1N (or other capital), 1(, the digits 4 to 8, 91 to 94, and 95 to 98.

The general principles here adopted are (a) to divide the facet into one or more sub-facets (called by Ranganathan first, second, . . . *levels*); and (b) to divide such a sub-facet into *zones*, each of which is introduced by an 'empty' signpost. The symbols 1, 1–, 19, 1(and 9 are of this type. Even if an ordinal notation is adopted, these two devices can still be used if required to differentiate various hierarchies within a facet.

One method of forming a hierarchy within a facet is by 'subject device'—characteristics already used in other schedules are used to divide a term in a facet. This is achieved by the use of an intercalation symbol (Colon uses a bracket) followed by the symbol for the term drawn from another schedule. In the hierarchical notation of Colon a hierarchy so formed is put at the end of an array: see Asia in facet I, where 4(Q7) follows all the other divisions of 4. In other words, the intercalator is of higher ordinal value than significant digits used for simple terms. In ordinal notation, on the other hand, the symbols 415, 42, 43, etc., might already be used to represent terms coordinate with that represented by 4, so that the intercalator must be of lower ordinal value than significant digits for simple terms. If the symbols for simple terms do not end in the first digit of the base, this digit can be used for subject device.

(3) *Within a main class or subject field* the facets must be distinguished and, when used alone, arranged in a preferred order. Using either hierarchical or ordinal symbols within each facet, this can be achieved by introducing them by signposts. These can be wholly or partially 'empty' of meaning. In the Colon Classification both types are used: facet I is in fact introduced by a dot, which is semantically empty; facet II is, however, introduced by the main class symbol U, which by itself means Geography. The order of terms formed is as follows.

U.1.5	Water (second-level facet)
U.5	Europe (facet I)
U5	Political geography (facet II)

The use of wholly 'empty' signposts is not essential. For example the second dot in U.1.5 could stand for Physical Features, so that U.5. would represent Europe—Physical Features. The use of punctuation marks as significant symbols might be too unconventional, but the notation can be so designed that numerals or letters act as signposts. The only essential point is that the symbol should place the heading in its preferred position in the order.

(4) *Within the classification as a whole* the fields or main classes must be distinguished and arranged in a preferred order. In Colon each main class symbol is also a signpost for the primary facet in the class, as U above. This feature is not essential. If the primary facet is introduced by another signpost, the main class symbol can be an isolate preceding it in a subject. For example, if facet II were introduced by a comma, giving U,5 for Political Geography, the U simply represents Geography, and 'U,' represents Geographical Problems.

Alternatively, if we consider a field of knowledge to be described by the most extensive class of 'substances' within it (e.g. the World in this schedule), then we no longer need a main class number preceding the signpost. All that is necessary is that all the terms in the *substance* facet be given an appropriate ordinal symbol.

(5) *To form unique compound terms with a facet* it is necessary to lay down a standard order for the combination of terms. This is illustrated in facet I, where Britain and Empire are linked to give 1–56: the lower symbol is cited first. In this case the connecting dash is simply a signpost indicating that the following symbol is drawn from the same facet. However, the connecting symbol can represent a specific directed relation, e.g. 'the effect of'. It must then be laid down which term of the relation takes preference—e.g. the term affected. If the symbol used is the zero, then Biogeography—Effect of climate would be U30287.

(6) *To form unique compound terms within a field or main class*, it is necessary that the combination order of facets be laid down. For the Colon schedules given above, this order is: facet II, facet I, second-level facet. Thus we get U287.419M.2 for Climate—West China—Mountains.

(7) *To form compound terms between fields or main classes*, using terms of directed relation (phase relations), the order of combination must be established as in (5). Thus, using zero as an 'influencing relation', Geography influenced by history would be UOV.

(8) *Zones*. The Colon classification started with a simple notation for simple terms—the digits 1 to 9, used as decimal fractions. They formed a notational base of 9—i.e. allowing nine coordinate terms to be accommodated in an array. This simple arrangement has grown steadily more complicated: (a) to accommodate more than nine terms in an array, 9 is used as an 'octavising digit', (b) to accommodate new hierarchies, capital letters and signposts such as the hyphen and bracket are introduced. Notationally, the effect of these developments is to form a much longer base, thus: –, 1 to 8, 91 to 98, 991 to 998, . . . , 99A to 99Z, 9A, A to Z, (. In theory, both the hyphen and the bracket could each be used to introduce all the digits which lie between them in the series, and a still larger base could be formed by introducing further signposts.

Within this base each block of digits (1 to 8, 91 to 98, etc.) forms a *zone* or sector. Above we have seen these zones allocated to different hierarchies in a facet. More generally, however, we can also allocate them to (a) groups of terms in an array, according to their frequency of appearance in the literature, (b) different facets in a field, (c) different fields in the whole classification.

PREFERRED INDEXING ORDER

As a result of all the devices so far discussed, a whole series of simple and compound terms has been constructed:

U	Geography
UOV	Effect of history
U.1	World
U.1.5	Water
U.1–56	British empire
U.15	Countries bordering Indian ocean
U.195	Temperate zone
U.1N5	Supranational organisation created in 1950s
U.1(Q7)	Moslem countries
U.5	Europe
U.95	Indian ocean
U5	Political geography
U50287	Effect of climate
U50V	Effect of history
U5.5	Europe
U54	Military geography

It is necessary that these be put into a preferred indexing order (the one above being that of Colon). In the main, this order will already have been settled by the order chosen for the ten hierarchies of facet I, which used symbols in the following ordinal scale: 'dot, dash, 1 to 9, A to Z, bracket'. It is only necessary to place the zero at the beginning of this list.

The ordinal series 'dash, 1 to 9, A to Z, bracket' was established to distinguish four zones within a facet, allocating the various hierarchies to these zones in a preferred order. The dot and the zero, however, are connecting symbols used to form compound terms. Let us consider more generally the provision of devices which will enable us to insert a compound term at any required point in a series of simple terms, such as the following schedules:

	(*A*)		(*B*)
13	Lower extremities	2	Morphology
132	Foot	3	Physiology
133	Ankle	4	Disease
134	Leg	412	Atrophy
135	Knee	etc.	etc.
136	Abdomen	83	Wounds
etc.	etc.		
82	Bones		
83	Muscles		

A subject such as 'Muscles of the lower extremities' is preferably placed between 13 and 132, before the individual parts of the body are discussed. But the symbol for Muscles cannot simply be justaposed to 13, for this leads to 1383, which is wrongly placed. Any symbol beginning with 131 would put the subject into its preferred position—thus, 13183. The digit 1 might therefore be used as a signpost to insert compound terms. But in that case, how would we insert into column (A) another simple term, such as Toes, which should also be between 13 and 132? Every number beginning with 131 would now be reserved for a possible compound term, and if, say, 1315 were chosen for Toes, the compound 13183 would be wrongly placed. It is clearly not possible to use the digit 1 to insert both simple and compound terms, i.e. as both a significant digit and as a signpost. Compound terms must be so symbolised as to precede all possible inserted simple terms. This result can only be achieved by the use of a connector digit lower in the ordinal scale than any significant symbol—for example, the dash. Thus, 'Muscles of the lower extremities' could be represented by 13–83.

In this subject the combined simple terms are both drawn from the same schedule (A), and the hyphen is thus an intra-facet connector. A further problem arises if one component of the compound term is drawn from another schedule or facet, (B)—what Ranganathan calls 'lamination'. A subject such as 'Wounds of the lower extremities' also combines two symbols 13 and 83. They cannot be linked by the dash, since 13–83 has already been used for a single-facet compound. Preferably, the new subject should be placed before 'Muscles of the lower extremities', so that the connecting signpost must be lower in ordinal value than the dash. If there are a number of other facets, each of which may require insertion into (A), then (a) each facet will need its own distinctive signpost, (b) all these must be lower in the ordinal scale than that used to insert single-facet compound terms, and (c) it is necessary to establish a helpful ordinal order among them.

The problem of inserting compound terms at preferred positions in a catalogue may be generalised a little further. The essential point is that, if both simple and compound terms are to be inserted after a certain term in a series, then the first new digit (i.e. the digit which is added to the symbol for the existing term) must, for compound terms, be lower in the ordinal scale than any new digit used for simple terms. This result may be achieved in two ways: (a) creating a new digit of lower value than the lowest used in the base for simple terms (i.e. using a 'signpost' as a connecting symbol), or (b) leaving a certain block of the lower-valued digits in the base free for the formation of compound terms. These two methods are illustrated below.

(1)	(2)		
B	B	Compounds containing 1 halogen	
BA	BC	Halogen =	iodine
BB	BD		bromine
BC	BE		chlorine
BD	BF		fluorine
etc.	etc.	etc.	

K	K	Carboxylic acids
KA	KL	Acetic
KB	KM	Propionic
KC	KN	Butyric

In column (1) all the digits in the alphabetic base (A, B, C, etc.) have been used to insert simple terms after B and K. In order to insert after K the compound term 'Monochloro-carboxylic acids', we cannot use the lowest ordinal digit (A), for that would give KABC—wrongly placed after KA. We must therefore create a new digit, such as the dash, which is arbitrarily given a lower ordinal value than A. The compound term K–BC is then placed before KA.

In column (2), for inserting simple terms, the added digit is always higher in value than the digit to which it is added (e.g. BA, BB, and KA to KK are not used). The lower-valued digits are thus reserved for the insertion of compound terms. It is then always possible to divide a later simple term (K) directly by an earlier (BD), to form a compound term (KBD) which will be placed before the next simple term KL. The effect of this device, 'retroactive notation', is that the average length of simple terms tends to be increased (since certain combinations of digits are excluded), but the need for a connecting digit for compounds formed from two terms in the series is avoided, giving a simpler symbol which in this case is also shorter.

The general result of this analysis is that symbols must be chosen in the following ordinal sequence: (1) signposts for facets, (2) connectors for combining terms within a facet and (3) significant digits within a facet. As usual, signposts can themselves be partially significant.

There is next the problem of the ordinal value of symbols of phase relation, such as the zero used by Ranganathan. Above, it has been used to represent a relation between two terms in one facet, and between terms in different main classes. The same relation can also hold between terms in two different facets of the same main class. Consider the following schedule:

(S)	Soil science
9	Kinds of soil
8	Structure
7	Constituents
6	Properties
6s	Salinity
5	Processes
5I	Infiltration
5J	Evaporation
4	Operations
4F	Irrigation

Here (S) is a main class, 9 to 4 are facet signposts (partially significant), s, I, etc. are terms in the facets.

Let us represent the relation 'the effect of' by OA, OB, OC. Within a facet, we have Infiltration—Effect of evaporation: (S)51OAJ. Between facets, we have Salinity—Effect of irrigation: (S)6sOB4F. Between main classes, we have Soil Science—Effect of War: (S)OC(W), putting (W)=War. The points we have to determine are these: (a) do we need to represent these relations by three different symbols, as here, or can we use a single symbol? (b) what are the correct ordinal values of such symbols—relative to the sequence already demonstrated, and relative to each other if more than one is needed?

The answer to the first question seems to be negative. The compound terms will be already divided into three groups, according to the origin of the second term, since the digits following the sign of relation (letter J, signpost 6, or main class bracket) have different ordinal values. We can therefore drop A, B, C, and use zero alone as the relation sign. What should be its ordinal value? Ranganathan places it below facet signposts, and provisionally we may agree with this.

The example just given, however, raises a further issue—the ordinal value of main class symbols. If these symbols are also used as signposts to the primary facet, then they should be the highest-valued of these signpost symbols. However, if they do not serve this function, but—as in the example—simply represent the main class they should I think be lower in ordinal value than any signposts. The most general 'effects' would then be placed first in shelf order.

Last, we come to common sub-divisions of literary form. As is usual with Colon, the UDC and Bliss, the symbols for these should precede all the others here mentioned. Our final preferred order of symbols is therefore as follows: (1) literary form symbols, (2) phase relation symbols, (3) main class symbols (if not used as signposts), (4) signposts for facets, (5) main class symbols (if used as signposts for primary facets), (6) connectors for inner-facet compounds, (7) intercalator (for ordinal notation), (8) digits for simple terms, (9) intercalator (for hierarchical notation).

CAPACITY FOR GROWTH

The outstanding characteristic of knowledge is its ceaseless development and change. It is added to, it becomes more detailed, more profound, wider in scope, and more intricately interconnected. The notation of a classification must allow of the interpolation of new terms and groups of terms at any point in the schedules. To illustrate the possible modes of change, let us repeat the Geography schedule, given earlier, in schematic form:

Main class U
  Facet I
    Hierarchy (*a*)
      Term 1
      Term 2
        Term 3
      Term 4
    Hierarchy (*b*)
  Facet II
    Term 5
Main class V
  Facet I
    Term 6

This structure can be modified in the following ways.

(1) *New term in array.* It may be required to add a new term to hierarchy (a) in facet I of main class U. The term may be added after term 4 (extrapolation) or interpolated between, say, terms 1 and 2. With a purely ordinal notation, this is always possible if one condition is fulfilled: no term must end with the first digit in the base used for significant symbols. This precaution is often neglected by Bliss: for example, his CYA, Fuels, immediately follows CY, and no new term can be interpolated between them.

With strictly hierarchical notation, interpolation in array is only possible if a gap has been left in the numbering. Ranganathan sometimes tries to allow for this by allocating digits to coordinate classes according to their rank in his list of 'seminal' terms (Appendix A). Extrapolation is possible if not all the digits in the base have been used. Ranganathan has also introduced the 'octave device' already mentioned to provide for unlimited extrapolation. The symbols 1, 2, 3, . . . 8, 91, 92, . . . 98, 991, 992, etc., are all taken to fall in the same array, the 'octavising digit' 9 being 'empty'. There is thus only one significant digit in each symbol.

Farradane has put forward another device for interpolation and extrapolation. For his original divisions he would use, say, capital letters from B to Y. For addition he would use digits of another species. Thus A, A1, A2, . . . A9, B, C, C1, C2, . . . C9, D, . . . Y, Y1, Y2, . . . Y9 are all taken to fall in the same array, the hierarchical aspect of the notation consisting in the fact that each symbol contains only one digit of the original species.

A special case is the insertion into one hierarchy of terms drawn from another facet. This is achieved in the hierarchical notation of the UDC by using the device, 'Subdivide like . . .'. Thus 542.9,

Other chemical processes, is subdivided like 66.09, Chemical processes in technology. The final 9 to 542.9 is here an 'octavising digit'. Ranganathan has used a bracket to perform this function. Ordinal notation could use a similar device.

(2) *New term in chain.* It may be necessary to insert a new term in the chain between terms 2 and 3. This is always possible for ordinal notation, and never possible for normal hierarchical notation. Once again, Farradane offers a solution based on the use of another species of digit. For a term inserted in the chain between B and its sub-class BG he would write Bg. For a term inserted between B and its divisions (potentially BA1 to BY9), he would write BΣ. More complex interpolation requires the use of further Greek letters.

(3) *New hierarchy.* It may be necessary to introduce a new hierarchy before (a), after (b), or between them. Once again, this is always possible for ordinal notation. For hierarchical notation, Farradane's method of putting new terms into an array is also applicable. Ranganathan has used various devices: (i) to insert a new hierarchy before the main hierarchy (digit 2 to 8) of the Geographical Divisions facet, he used an empty digit of ordinal value lower than 2 (namely, 1–); (ii) to insert a new hierarchy between 1– and 2, he used a further empty digit (forming the signpost 19); (iii) to insert further new hierarchies between 19 and 2, he introduced two new species of digits—the capital letters, and the bracket, both of ordinal value greater than 9.

(4) *New facet.* The insertion of a new facet into a field or main class is clearly quite a fundamental revision of the class, and may require a reorganisation of its notation. The existing facets may be 'signposted' or 'retroactive'. For the first style, one method of inserting a new facet is to provide a new signpost digit of the appropriate ordinal value. Alternatively, the new facet can be added as a new hierarchy to the existing facet which precedes it in citation order. In this case, a term in the new facet is linked to one in the existing facet by the auto-bias signpost used for combining terms within a facet. If the retroactive notation is ordinal the insertion of a new facet is always possible, and if it is hierarchical the devices of Farradane and Ranganathan, mentioned in (1) and (3), are applicable.

(5) *New main class.* Hierarchical notation, in which all main classes are represented by symbols of equal length, can insert new main classes by the introduction of new digits. For example, Ranganathan used Greek letters to interpolate in the Roman alphabet, giving them the appropriate ordinal value. Farradane's method in (1) is also applicable. Ordinal notation offers no difficulty.

BREVITY AND SIMPLICITY

The main features of notation have been summed up by Berwick Sayers: 'It may consist of any symbols that are capable of marking all parts of the scheme. It should, however, be brief, simple, flexible, and mnemonic.' Flexibility, he continued, means that 'the notation symbols must be capable of expansion to mark insertion without dislocating the rest of the notation. This is the cardinal requirement of notation.' Mnemonics cannot be used *systematically* in a purely 'ordinal' notation, and will not be discussed here.

Flexibility (maintenance of helpful order no matter how the pattern of knowledge develops) is not only the cardinal virtue but also the whole purpose of notation, and has already been considered. It is implicit in all that follows. The problem we must examine is how to achieve flexibility with maximum brevity and simplicity. It is well known that these two latter criteria conflict with each other. The simpler the notation, i.e. the fewer types of digit it employs, then the fewer are the combinations which symbols of a given length can provide—and, hence, the longer the symbols must be made to accommodate a given number of terms. The advantage of brevity in the card catalogue or on documents is self-evident. The advantage of simplicity is, of course, that all the digits are familiar, and their ordinal order can also be a familiar conventional one (that of the numerals or the alphabet).

Dewey used only numerals, but recognised that since there are more letters of the alphabet than numerals, use of the former would give shorter class symbols. Indeed, though he adopted 'the simplest numerical scheme', he stated that 'my strong leaning to the shortest of all methods leads me to find all possible advantages in the letter method'. Bliss was also attracted by the larger base which letters provide, and argued persuasively that 'a notation of letters or of letters and figures combined is virtually as legible as a notation of figures alone, and it is shorter to write and is more readily remembered'.

Bliss used numerals for common sub-divisions, and an alphabetical base distributed throughout the whole set of simple terms in his main schedules—for example:

C	Chemistry	F	Botany
CG	Analytical	FD	Physiology
CGQ	Quantitative	FDE	Metabolism
CI	Inorganic	FK	Systematic
CIU	Phosphorus	FL	Cryptogamia
CO	Organic	FLF	Algae

I will call this an 'enumerative' or 'fenced' style of notation, since in order to combine terms it is necessary to separate them by a 'fence'. Ranganathan set out his schedules in a rather different way. Each facet is first of all symbolised by combining (a) a main class letter and (b) a facet indicator (a colon, comma or semi-colon). Simple terms are then represented by the use—repeated in each facet—of the numerals. We may call this a 'labelled' style of notation, since each main class is labelled, thus:

E	Chemistry	I	Botany
E : 3	Analytical	I : 3	Physiology
E : 34	Quantitative	I : 33	Metabolism
E1	Inorganic	I1	Cryptogamia
E151	Phosphorus	I22	Algae
E5	Organic		

Dewey, Bliss and Ranganathan have all urged that notation should be as brief as is feasible. Ranganathan has demonstrated that on average his notational symbols for a given subject are shorter than those of the Decimal Classification, and his symbols could no doubt be compared with those of Bliss. But there are several different features of any notation which contribute to the length of the resulting symbol. As well as the differences in structure just noted, I have also mentioned that the notation of Bliss is largely ordinal, and Ranganathan's is largely hierarchical. Though general comparisons do quite fairly pronounce on the combined effect on length of all such features, they do not analyse the contribution of each feature. Consequently, they give little guidance as to the best notation to use in constructing a new classification.

To obtain a quantitative understanding of the relationships between brevity of symbol and style of notation, it is necessary to construct a series of model notations in which only one feature is varied at a time. For example, E. J. Coates has fitted the same set of schedules with both ordinal and hierarchical notation: the average length of symbol for simple terms in the first was 66% of their length in the latter.

The relative simplicities of ordinal and hierarchical notation are illustrated by the following schedule:

Hierarchical		*Ordinal*	*Schedule*
Ranganathan	Farradane		
1	B	B	World
1–56	*BA1*	*BAB*	*British empire*
11	BB	BB	Eastern hemisphere
12	BW	YC	Southern hemisphere
191	*BW1*	*YCB*	*Equatorial zone*
1N	*BW5*	*YCM*	*League of Nations*

7	BX	YD	America
71	*BXb*	*YDB*	*North*
72	BXB	YE	Canada
728	BXBY	YZ	Quebec
7291	*BXBY1*	*YZB*	*Yukon*
79	*BX5*	*YZM*	*New Zealand*
8	BY	Z	Australia
91	*BY1*	*ZZ*	*Land within ocean*
12(16)	16	12	original no. of digits
31(39)	39	32	total no. of digits

The terms not italicised represent an original schedule to which italicised additions have been made. The term 'New Zealand' has been interpolated in an array between America and Australia; the array containing Quebec has been extrapolated by the addition of 'Yukon'; 'North' has been interpolated in the chain between America and Canada; and four new hierarchies have been included. The first column of notations is approximately as Ranganathan has designed it; the second is drawn up along Farradane's lines; and the third is purely ordinal. All three are so arranged that there are no convenient gaps to accommodate new terms.

We may note first that the ordinal notation is the briefest: though there are only 12 digits for the original schedule in Ranganathan's symbolism, this brevity is achieved by an 'ordinal' device—7 and 8 are in fact subordinate concepts to 1, and a purely hierarchical notation would indicate this, making a total of 16 digits for the original schedule.

The original symbols of Ranganathan's notation are 1, 11, 12, 7, 72, 728, 8: all of one species. To make the additions, Ranganathan resorts to two devices: (1) he departs from hierarchical notation at 71 and 79, since he has no method of interpolation in chain or array; and (2) he introduces new digits, the dash and the capitals. Farradane's original symbols are B, BB, BW, BX, BXB, BXBY, BY: again all of the same species. To accommodate the new terms, he introduces new digits, the numerals and the small letters. Ordinal notation accommodates them by the further use of digits of the original species—capitals—leaving to the words of the schedule the task of indicating the relations between terms. It is clear, therefore, that for use within a facet, ordinal notation is both briefer and simpler than hierarchical.

Proceeding in similar ways, we can examine a series of alternative notational devices as to their brevity and simplicity. I have published in the *Journal of Documentation* comparisons of (1) hierarchical symbols and semantic compounds, (2) signposted and retroactive notation and (3) enumerative and labelled notation. The following table illustrates some conclusions reached in these studies:

	(*a*)	(*b*)	(*c*)	(*d*)
Chemistry (main class)	54	54	56	C
Organic chemistry (facet I)	547	547	567	7
Saturated acyclic	5472	4	5677	72
Subst. hydrogen	54721	3	56778	721
4 carbons	547214	23	567789	7214
Branched	5472142	12	5677899	72142
Physical chemistry (facet II)	541	541	566	—
Valency	5415	5415	5666	5
Bonds	54157	54157	56667	57
Principal	541571	54158	566678	571
Simple	5415711	54159	5666789	5711
	(*e*)	(*f*)	(*g*)	(*h*)

Columns (*a*) and (*b*) for Organic Chemistry illustrate the use of (*a*) hierarchical symbols and (*b*) compounds of isolates (4, 3, 23 and 12) to represent the attributes in a facet. The substance isobutane is represented by the hierarchical symbol 5472142 or by a 'retroactive' compound 547.432312. The general conclusion is that the use of the latter method, using the same notational base of digits, will always lengthen the term symbols, as here, and may do so considerably, though it allows of a far more flexible combination of attributes (see my paper III).

Columns (*e*) and (*f*) for Physical Chemistry contrast the use of (*e*) hierarchical flexions and (*f*) ordinal isolates. In the former the hierarchy of terms in the verbal headings is matched in the notation; in the latter it is not. As is to be expected, the ordinal symbols are shorter (see my paper III).

The facets in the two columns (*a*) and (*e*) can only be combined if a connective symbol is used to link them—they are an example of signposted notation. Thus 'the simple bonds in isobutane' is represented as 5472142 : 5415711. The two numbers could not be run together without confusion. The two columns (*c*) and (*g*) show symbols which could be used in retroactive notation, without a signpost—the compound symbol would be 56778995666789, the point of linkage being indicated by the 'backwards' movement from 9 to 5. The relations between symbol length and notational style are here rather complex, but in general terms, it is found that unless a large number of signposts are used the retroactive notation is shorter (see my paper V).

Last, both these sets of columns are examples of an enumerative or fenced notation, in which the notational base is distributed throughout the whole of the classification schedules. The effect of this is that the class number giving rise to facets (in the first column it is 54, Chemistry) is repeated in every facet, as in the compound 5472142 : 5415711. Columns (*d*) and (*h*) represent what

I have called a labelled notation. The main class number is removed from the facet numbers, and is symbolised by a digit of another, species, C. The first facet needs no signpost (7, Organic Chemistry, is here a term in the *substance* facet), and the second facet is introduced by a dash. The above compound is now represented as C72142–5711. In this example the labelled notation is briefer than the enumerative. This is true in general, though the quantitative relations are somewhat complicated (see my paper II).

ALLOCATION OF ORDINAL SYMBOLS

In hierarchical notation the formation of flexional symbols is governed by the hierarchical relations among the concepts symbolised. In ordinal notation no such restriction is placed upon the symbolism—we are free to allocate symbols to terms in whatever way will achieve maximum brevity.

There are four sets of symbols used in a faceted classification: for main class names, for simple terms, for literary form divisions and for connecting symbols (facet indicators, phase relations, inner-facet connector, intercalator). Each symbol consists of one or more digits of the notational base. With ordinal notation we can decide quite arbitrarily how many digits each particular symbol is to have. Is there any principle upon which such a decision can be based?

This problem has been considered mathematically in information theory. Brillouin, for example, considers encoded information consisting of discrete symbols, each of which has a certain 'cost'—which we may here interpret as 'length', n. There are N_1 symbols of length n_1, N_2 of length n_2, and so on up to n_m, the maximum symbol length in use. The total length of the encoded 'message' is $N_1n_1 + N_2n_2 + \ldots = T$. The total number of symbols in the 'message' is $N_1 + N_2 + \ldots = G$. Symbols of any given length, say n_x, appear with a probability $p_x = N_x/G$. From these expressions Brillouin shows that the total number, P, of different 'messages' of length T that we can construct by interchanging the positions of different symbols is given by $\ln P = -G\Sigma p_x \ln p_x$. The most economical coding system is one which makes P a maximum, and this condition is satisfied if $\ln p_x = -An_x$, where A is a constant. In other words, maximum brevity is obtained if the length of a symbol, n_x, is inversely related to the probability of its occurrence (actually, proportional to $-\ln p_x$).

This condition is applicable to notational symbols for classification: symbols should be so allocated that the most frequently used are the shortest, with length increasing logarithmically as frequency declines. Such fine quantitative adjustment of symbol length to frequency of occurrence has rarely been attempted in coding systems, because of lack of detailed information as to term frequencies—though statistical studies have shown that *natural* languages do approximately conform to the mathematical relationships involved.

In a general way, however, the rule has been followed in classification. Hierarchical notation, for example, assumes that the more specific is a subject, the less frequently it will occur, and so adjusts length of symbol to correspond exactly to the order of a term in the conceptual hierarchy. Again, in Ranganathan's 'octave notation' for an array, the first octave (1, 2, 3, . . . 8) is allocated to those terms with the greatest 'literary warrant', and later octaves with longer symbols (91 to 98, 991 to 998, etc.) to terms appearing less frequently.

Statistical evidence of literary warrant is rarely available. In one particular field, however, a quantitative analysis has been made. At the Chemical-Biological Coordination Center in Washington, over 40 000 chemicals were encoded in a punched card file in the years up to 1953. The coding was by functional group—benzene ring, keto-group, nitro-group, and so on—and Dale and Heumann have analysed their frequency of occurrence. For example, in 53.61% of the compounds there was a non-cycle single carbon atom; in 34·73%, a single benzene ring; in 13·35%, a tertiary amine. Again, the frequencies of the simple anions of chlorine, bromine and iodine were 13·68, 1·49 and 1·18%. Detailed studies of this kind would be of great value in the economical allocation of symbols.

CHOICE OF BASE

Some consideration has already been given to the choice of notational base—the set of digits used in constructing symbols. Brevity of symbol is favoured by the use of a large base, which necessarily includes more than one species of digit—simplicity is lost. This may detract from ease of working: the digits and their ordinal values may be unfamiliar. On the other hand, the use of one species of digit for simple terms, and of other species for connecting symbols, means that the 'make-up' of a compound symbol is readily apparent, and this is claimed to add to the ease of working with a notation.

The species of digit used in the main schemes of classification are (1) numerals, used in all schemes; (2) capital letters, used in all schemes except the original Dewey; (3) small letters, used by Ranganathan, Bliss and (for alphabetical device) the UDC, (4) punctuation marks, used particularly by Ranganathan and the UDC; (5) mathematical and other symbols, used by the UDC and Bliss; and (6) Greek letters, used only by Ranganathan but suggested also by Farradane.

In all the main schemes each digit is used independently, but recently there have been suggestions to combine two digits to form a compound which is in effect a new species of digit. Thus J. E. Daily proposes the formation of a very large base in which the combinations *A to *Z, *1 to *9, ″A to ″Z, ″1 to ″9, :A to :Z, :1 to :9, 1A to 1Z, 2A to 2Z, . . . 9A to 9Z would form 15 new species. Together with 1 to 9 and A to Z, this would give a base of 464 digits, plus certain additional digits (the dot, the dash) used as connecting symbols. The uses of 'zones' in an array by Ranganathan, discussed earlier, similarly creates a large base. At the other extreme, the pronounceable alphabetic notation of de Grolier uses a strictly limited base—the twelve capitals A, E, I, K, L, M, N, O, P, S, T and U, and five mathematical signs as connectives.

There are strong arguments against a restricted base. If we are using a hierarchical notation, in which the chains and arrays of symbols reflect the hierarchy of terms, then we cannot be content with a notation which allows of only a few divisions of any class. Nature is not bound by such low figures. On the other hand, if we choose to use an ordinal notation, which does not mirror the hierarchy of terms, the need for a brief notation calls for as large a base as is compatible with simplicity of notation.

In the last analysis the choice of a notational base will be subjective. Even a large, complex base can become familiar in time, yet it is certainly a barrier for the newcomer to a classification. As shown earlier in this chapter, the insertion of new terms, hierarchies and facets into a fully occupied hierarchical notation will demand the introduction of new digits and new species of digits, whereas ordinal notation can accommodate such additions without extending its base. Consequently, if we adopt a hierarchical notation, it will be as well not to use all the appropriate species of digits in our 'first edition' but to keep some in reserve for later expansion. An ordinal notation need not act so cautiously. Again, a 'labelled' notation, if it uses one species of digit (e.g. the capitals in Colon) simply to label main classes, is perhaps restricting its possible ramification base.

It has been argued by E. J. Coates that, after readily recognised ordinal relations, the next most important feature of a notation is that it should be easy to retain mentally. This is achieved partly by brevity and partly by simplicity of symbol. A symbol containing digits of only one species (e.g. 4726214) may be easier to retain than one containing several species (e.g. 4U2k21P), and easier still may be one with a single 'separator' (e.g. 472k214). Again, some 'separators' may be of greater psychological value than others—a character used in normal writing as a 'separator' may be more readily retained (e.g. 472–214). The International Study Conference on Classification for Information Retrieval urged the need for psychological tests as to the relative ease of retention of various symbolisms.

The value of pronounceability in aiding ease of retention has been stressed by the French school of classifiers. Adapting de Grolier's work, I have published in the *Journal of Documentation* the details of a pronounceable retroactive ordinal notation that provides 718 syllables of no more than three letters, which can be freely combined to form pronounceable symbols such as 'bipbes', 'letfyn', 'nosgro'. Further syllables of more than three letters can be formed if desired (see my paper VI).

SUMMARY OF PRINCIPLES

The preceding survey will have left the reader with the impression of a great variety of notational devices and alternative methods of achieving the one end of mechanising order. In fact, one reason for the survey has been to create such an impression. The notational set of tools which modern classification has developed is indeed very varied. This description of its flexibility and intricacy may serve to drive home two points. (1) There are no hard and fast rules whereby a given problem of order must be solved by only one specific symbolism: there are always alternative methods available. (2) Notation is only a tool; it must always be subservient to the indexing order it is designed to preserve and display; it must never be allowed to dictate that order.

It is important to emphasise the second point. Two of the major schemes of classification, the Decimal and the Colon, have been named after their leading notational feature, and Ranganathan in particular has sometimes written of the development of classification as though it were simply a matter of improved notation: 'the whole history of classification is a history of breaking rigidity'. The consequence of this emphasis has been that 'any change of notation from numerals to alphabets, from alphabets to numerals, and so forth was regarded as a new classification system. Rather later, embodiment of different or novel clerical media and operations, such as edge-notched cards and magnetic recording, were apt to be so regarded. Today we accept, perhaps, these clerical factors as links in the chain, not the chain itself. Even if they are good, they cannot save a poor system though, if poor, they may spoil a good one.' These words of Fairthorne echo those of Bliss quoted earlier: 'Notation, however real its service, does not make the classification, though it may mar it.'

The essence of the matter is this. The only purpose of classification is to put subjects into a helpful order. What constitutes this order can only be decided by a study of the subjects themselves, not by a study of the symbols by which we represent subjects. We must make our preferred arrangement before we can mechanise it. Notation must follow subject analysis. We must use symbols as Humpty Dumpty used words, 'so that they mean just what we choose them to mean, neither more nor less'.

The second reason for the preceding analysis of notation has been to make clear exactly what it is that symbols do mean, so that in using them we may understand precisely what role they are playing in the notation. The third reason is that we might deduce from the survey what general principles are applicable to all forms of notation for the classified catalogue if it is to fulfil its purpose.

The survey has shown that the three purposes of notation can be fulfilled as follows: (1) It will *name* the subject of the reference if there is provided an index to its symbols, in a language familiar to the searcher. (2) It will allow the subject to be *located* if its symbols are constructed from a relatively small number of digits whose relative order is familiar to the searcher. (3) Notation can reveal relations between subjects in two ways: (a) by juxtaposing their respective symbols and (b) by representing the subjects by symbols of similar pattern—i.e. the symbols of related subjects have certain elements in common. A search for related subjects then becomes a search for symbols which have similar patterns. The patterns used can be semantic compounds, in which the component elements are independent and retain their meanings when separated (e.g. caveman, 547 : 541); or they can be flexional symbols, the elements of which have different meanings in different contexts (as in the traditional decimal symbol).

In a semantic compound the two or more elements are often linked by a connective symbol. This may be simply a fence, separating successive elements of the compound; or it may be a signpost, indicating the schedule from which the following element is drawn; or it may itself have a distinctive meaning, expressing the nature of the relation between the terms it links.

Within a semantic compound, the elements can be either flexional hierarchical symbols (as in 547 : 541) or ordinal isolates (as in cave-man). The main use of flexional symbols is to display, in the notation, the hierarchical relations between indexing terms. This feature can, however, be displayed by the verbal headings of a schedule or catalogue, and the notation can be purely one of ordinal isolates, serving only to mechanise a preferred order.

The ordering of classification symbols is effected by two mechanisms: (a) in general, by using an ordinal series of digits which form decimal fractions (more generally, radix fractions), so that new decimals may be inserted at any point without disturbing the relative order of the terms; (b) by including in this ordinal series digits which, as fences, signposts or relational isolates, serve to insert compound terms at preferred points. A variety of devices is available for compound formation, and the only essential feature is that the digits used should form the following ordinal series: (1) literary form symbols, (2) phase relation symbols, (3) main class symbols (if not used as signposts), (4) signposts for facets, (5) main class symbols (if used as signposts for primary facets), (6) signposts for inner-facet compounds, (7) digits for interpolation in hierarchical chain, (8) intercalator (for ordinal notation), (9) digits for simple terms, (10) intercalator (for hierarchical notation), (11) digits for interpolation in hierarchical array.

An ordinal notation, which uses only isolates, allows of unlimited interpolation of new terms in array and in chain, and of new facets. A hierarchical notation using flexional symbols can only achieve this flexibility by introducing new 'non-significant' digits, such as the 9 in Ranganathan's 'octave device', the bracket in his 'subject device' or the new species of digits introduced in Farradane's notation.

All the above principles relate to the flexibility of the notation—its capacity to place headings in a preferred order, to reveal relations between headings, to insert compound terms where required, and to allow for growth of the schedules. This flexibility must be achieved with maximum brevity and simplicity of symbol. Given a certain notational base (i.e. a certain level of simplicity), various factors contribute to the average length of the resulting simple and compound symbols.

An analysis of these factors suggests the use of a notation in which (1) main classes are labelled by distinctive signposts, e.g. digits of a species other than that used for simple terms; (2) within a main class, retroactive notation is used rather than facet signposts unless the number of facets is large; (3) within each facet, simple rather than compound symbols are used unless there is the need to represent a very large number of terms compounded of relatively few characteristic attributes (as with chemical substances, for example); and (4) the simple symbols so used are purely ordinal, not matching the hierarchical relations of the verbal headings.

BIBLIOGRAPHY

H. E. BLISS. *Bibliographic Classification.* New York, 2nd ed., 1952

H. E. BLISS. *Organization of Knowledge in Libraries.* New York, 2nd ed., 1939

L. BRILLOUIN. *Science and Information Theory.* New York, 1956

Classification Décimale Universelle. *Tables abrégées, principes et regles de la classification.* Brussels and Paris, 1947

E. J. COATES. 'Ordinal and hierarchical notation'. *Classification Research Gp. Bull.*, 1956, No. 1, 15–19

E. J. COATES. 'Notation in classification'. *ibid.*, 1957, No. 2, D1–D19 (reprinted in *Proceedings of International Study Conference on Classification for Information Retrieval, London, 1957*)

G. CORDONNIER. 'Méthodes nouvelles de documentation', *CNOF*, 1951, **25,** Nos. 4–6

J. E. DAILY. 'A notation for subject retrieval files'. *Amer. Docum.*, 1956, **7,** 210–14

E. DALE and K. F. HEUMANN. *Statistical Information on Component Parts of Chemical Compounds.* Washington, Chemical-Biological Coordination Center, 1955

E. DE GROLIER. 'L'étude des problèmes de classification'. *Rev. Docum.*, 1953, **20,** 105–16

E. DE GROLIER. 'Nouvelles recherches sur la symbolisation'. *Rev. Docum.*, 1956, **23,** 1956, 13–21

G. M. DYSON. 'The new notation for organic chemistry'. *Aslib Proc.*, 1949, **1,** 5–21

G. M. DYSON. 'Advances in classification'. *J. Docum.*, 1955, **11,** 12–18

R. A. FAIRTHORNE. 'Essentials for document retrieval'. *Spec. Libraries*, 1955, **46,** 340–53

R. A. FAIRTHORNE. 'Information theory and clerical systems'. *J. Docum.*, 1953, **9,** 101–16

R. A. FAIRTHORNE. 'Matching of operational languages in documentary systems'. *Royal Aircraft Establishment Library Memo. No. 27* (Farnborough, Hants, 1956)

J. E. L. FARRADANE. 'A Scientific theory of classification and indexing'. *J. Docum.*, 1952, **8,** 73–92

D. J. FOSKETT. 'Two notes on indexing techniques'. *J. Docum.*, 1962, **18,** 188–192

J. E. HOLMSTROM. 'The relation between referencing symbols and language'. *Rev. Docum.*, 1950, **17,** 20–27

J. MILLS. 'Library classification'. *J. Docum.*, 1970, **26,** 120–160 3rd ed., 1967

S. R. RANGANATHAN. *Prolegomena to Library Classification.* London, 3rd ed., 1967

S. R. RANGANATHAN. 'Self-perpetuating scheme of classification'. *J. Docum.*, 1949, **4,** 223–44

S. R. RANGANATHAN. 'Optional facets' (1) to (21), appearing in *Abgila*, 1949–53

P. A. RICHMOND. 'Simple nmemonic large-base number system for notational purposes'. *J. Docum.*, 1958, **14,** 208–11

W. C. BERWICK SAYERS. *Manual of Classification.* London, 3rd ed., 1955

Universal Decimal Classification, Abridged English edition. London, 1948

B. C. VICKERY. 'Notational symbols in classification'. *J. Docum.*, 1952, **8,** 14–32; part II: 'Notation as an ordering device'. *ibid.*, 1956, **12,** 73–87; part III: 'Further comparisons of brevity'. *ibid.*, 1957, **13,** 72–7; part IV: 'Ordinal values of symbols'. *ibid.*, 1958, **14,** 1–11; part V: 'Signposted and retroactive notation' and part VI: 'Pronounceable retroactive ordinal notation'. *ibid.*, 1959, **15,** 12–16

B. C. VICKERY. 'The changing structure of knowledge'. Abgila, 1954, **1,** 137–47

5

CLASSIFICATION IN INDEXING

The purposes of information indexing have been well expressed by Metcalfe. 'The purpose is not the specification or codification of documents for the sake of specification, in any supposed universal language or code. It is specific reference to and/or generic survey of information in its literary or documentary forms.' The degree of generic survey required depends on 'what we find or hope to find together in a useful togetherness, from only information on needles, say, to all information on articles made of metal, from a needle to an anchor.'

The assumptions on which this book is based were set out in the Preface to the first edition. I assume that an index to an assembly of subjects which are compound (ABCD, etc.) should provide: (1) specific reference to each subject; (2) generic survey in two forms —reference to ABCD from any of its component terms (A, B, etc.) and reference to each term from any other term which includes it; and (3) some degree of reference to each term from terms coordinate in array to it.

A fully flexible indexing system, capable of meeting all the demands of users, needs to have these capacities. For economy, or for special uses, some capacities may be sacrificed—particularly capacities for generic survey. But the target of modern information retrieval is the fully flexible system, and it is this that an index based on a faceted classification tries to provide.

THE ALPHABETICAL SUBJECT INDEX

For direct reference to specific subjects, there is no better tool than the alphabetical index. Access to each subject is directly by means of the words of a natural language—no transformation into a class or code symbol is necessary. There is freedom to introduce new terms whenever and wherever they are needed. However, this apparent verbal freedom has its dangers and it is often found necessary carefully to control the terms and relations used. This control can be based on facet analysis.

Alphabetical subject indexes use several types of heading. Some headings are single words—War, Religion, Boats, Colour, Acridines, and so on. Others are single words qualified by parentheses—e.g. Calculus (mathematics) and Calculus (medicine); Noses (anatomy) and Noses (aeronautics); Relaxation (medicine), Relaxation (engineering) and Relaxation (mathematics). The next type is the adjectival phrase—Bacteriostatic substances, Statistical analysis, Alimentary canal, Solar system, Nickel steels; a complex example is 'Acid Bessemer sulphurised carbon steels'. The prepositional phrase is also found—Fertilisation of flowers, Blenders for mycelium, Ionisation in water. These forms of heading are all in natural language, but formalised, non-grammatical headings are also used. The adjectival phrase is inverted, to bring the noun to the front, as in

Anatomy, morbid	Antibiotics, actinomycete
History, ancient	Blood, human, peripheral

The last type of alphabetical heading is the noun combination:

Coal—Production
Radiation—Detection—Instruments—Design
Leaves—Abscission—Retardation
Spectra—Atoms—Helium—Fine structure
Liquids—Structure—Acoustic wave study—Methylene chloride

In controlling the vocabulary of an alphabetical index, the first step is to restrict the number of types of heading used. Marie Prevost urged the use of noun combinations whenever possible. Prepositional phrases can usually be replaced by a noun combination (e.g. Fertilisation of flowers by Flowers—Fertilisation), but some have to be retained since the phrase as a whole has a specialised meaning (e.g. Bodies of revolution). Adjectival phrases cannot be avoided so readily, as they are needed to specify entities which have no individual name (e.g. Actinomycete antibiotics). Nevertheless, the target at which to aim is a coordinated succession of nouns, qualified nouns (adjectival phrases) or prepositional phrases acting as nouns.

THE ANALYSIS OF PHRASES

One of the first to explore these problems was the American librarian C. A. Cutter. In considering a compound name such as Agricultural Chemistry, Cutter was concerned to enquire, under which word in the compound would the user be expected to look if he sought this particular topic? The rules that he discussed were therefore concerned with specific entry, not with classing, and this has been the case with others who have put forward rules for inversion.

Cutter recognised three types of phrase that might be inverted: (a) the noun preceded by an adjective, such as Agricultural Chemistry; (b) the noun preceded by an adjectival noun, such as Death Penalty, and (c) the prepositional phrase, such as Fertilisation of Flowers. He discussed three possible rules: (1) never invert, (2) invert only type (a), and (3) invert any type of compound name if doing so brings to the front what we consider to be the 'most significant' word of the phrase.

The first rule would introduce no grouping other than that provided by natural language, e.g.:

Agricultural botany
chemistry
economics
history
implements
land
machinery
rents

The second rule, as far as classification is concerned, implies that the noun is the preferred basis of grouping, e.g.:

System, economic
educational
financial
social
solar

The third rule, recognising that both the others are too absolute, leaves inversion to the discretion of the indexer—in fact, it is no rule at all, and offers no classificatory guidance.

Coates has intensively analysed adjectival phrases according to the semantic categories of their components, and applied his principles in the construction of the alphabetical *British Technology Index*. He recognises four semantic categories into which words may be classed:

Thing—e.g. Beams, Inks, Buses, Engines
Material—e.g. Aluminium, Steel, Wood
Action—e.g. Moulding, Welding, Labelling, Collision
Property—e.g. Velocity, Thickness, Hardness, Fertility

Other words in phrases may represent the type or purpose of a thing, material, action or property. Binary combinations of categories give rise to a dozen kinds of compound name, and a rule as to the inversion of each is put forward by Coates:

Example	*Categorical analysis*	*Invert?*
Aluminium beams	Material type of thing	Yes
Research satellites	Action type of thing	Yes
Illustrated books	Property type of thing	Yes
Anodised aluminium	Property type of material	Yes
Injection moulding	Action type of action	Yes
Elastic collisions	Property type of action	Yes
Printing inks	Action purpose of thing	No
Rocket engines	Thing, part thing	No
Rocket research	Thing, action	No
Lightning fires	Action type of action	Yes
Hardness ratio	Property of property	No

Thus, as far as classification is concerned, the main effect of Coates' rules is to group by purpose, if present, otherwise by the 'thing' word, if one is present. The analogy to preferred groupings in schemes of faceted classification is evident.

When Coates does not invert, he introduces the colon as a separator between the words in the phrase:

Printing:	inks
Rockets:	engines
Rockets:	research
Hardness:	ratio

In form and in fact, the phrase becomes a heading and sub-heading. This opens the way to rules for compounds other than adjectival phrases, such as the following:

Example	*Categorical analysis*	*Invert?*
Welding by lasers	Action by thing	Yes
Thickness of beams	Property of thing	Yes
Labelling for packaging	Action for action	Yes

A further step is to develop rules for compounds of more than two words, e.g. 'Welding of foil by lasers' is analysed as 'action on thing by thing', and the heading formed is Foil: Welding: Lasers. To sum up, the grouping found in the *British Technology Index* is controlled by the classing of terms into categories and a pre-determined combination sequence of categories, in a way that is completely analogous to the principles used in faceted classification.

NOUN ORDER

The next problem is the choice of the order of nouns in a compound heading, and various rules have been put forward for this purpose. Kaiser separated terms into two types: *Concretes*—substances, equipment, things, forms of energy; and *Processes*—static or dynamic conditions of the Concrete. He constructed headings by putting the Concrete first, e.g.:

Calico—Printing
Coal—Analysis
Rabbit—Poisoning
Liver—Injury

Any word which expressed a combination of Concrete and Process was replaced, e.g.:

Bibliography = Book—Description
Agriculture = Land—Cultivation

This rule groups terms into two categories—Concrete, or what I have called 'thing, substance, entity' in Chapter 2, and Process, which covers all the attributes, relations and operations listed on p. 21. It is analogous to Ranganathan's distinction between Personality and Energy, described in Appendix A.

As it stands, Kaiser's rule for noun order is applicable only to compounds containing two nouns—one Concrete and one Process. Rules for controlling the order of longer compounds can be derived from facet citation chains such as those diagrammed on p. 30. Nouns can be combined in such orders as:

Thing—part—constituent—property—measure
Thing, property or process—operation—agent

Some examples of such headings are:

Antibiotics (thing)—Production (operation)—by Soil organisms (agent)
Plants (thing)—Disease (process)—Control (operation)—by Antibiotics (agent)
Liquids (thing)—Tensile strength (property)—Measurement (operation)

Between each pair of successive categories in such a citation chain there exists a directed relation, and Farradane has suggested the use of a series of directed relations to control noun order. Three of the most useful of his relations or 'operators' are those of 'reaction', 'causation' and 'appurtenance'. Each operator is represented by the combination of an oblique stroke and another punctuation mark, the stroke being placed adjacent to the term which comes first in combination order. Some examples follow:

'Reaction' operator
  Agent precedes Object of action, e.g.
    Morphine/-Respiratory centre
  Action precedes Object of action, e.g.
    Hydrolysis/-Sucrose
  Process precedes Product, e.g.
    Production/-Alloys
'Causation' operator
  Raw material precedes Product, e.g.
    Sucrose/: Glucose (formed from sucrose)
  Agent precedes Action, e.g.
    Acid/: Hydrolysis
'Appurtenance' relation
  Object precedes Constituent, e.g.
    Steel/(Chromium
  Object precedes Property, e.g.
    Cobalt/(Elasticity
  Property precedes Measure, e.g.
    Elasticity/(Young's modulus

Some complex examples of the use of these operators are:

Gases/(Reactions/(Kinetics
Acid/: Hydrolysis/-Sucrose/: Glucose
Cat/(Eye[-/Atropine]/(Iris/(Dilatation

In the last example Atropine/-Eye is the preferred directed order (Agent precedes object of action). Since this conflicts with the order Cat/(Eye, which is given precedence, Atropine is placed after Eye in brackets.

The noun order arrived at by the use of Farradane's operators is not always the same as that of the facet citation chains described earlier: Farradane uses the order Agent, Action, Raw material, Product, whereas on p. 32 I have suggested an exactly opposite combination order. Only practical experience can decide which is the more helpful to the user. The main point is that each system offers rules to achieve consistency in the construction of compound subject headings.

The principles of categorical analysis and relational operators have been combined and applied to alphabetical indexing by Austin and Butcher for the *British National Bibliography*. They started with a natural language phrase expressing a topic—for example:

(a) Plating the spokes of bicycle wheels
(b) Contamination of rivers by power station feedwater
(c) DDT for controlling aphid predation on roses

They first categorise each term as 'entity' or 'attribute'. An 'entity' is comparable to Coates's 'thing' or Kaiser's 'concrete': it is defined as 'a thing, whether concrete object or mental construction'. Grouping is first of all by entity. Attributes are the properties of entities, their activities, or the properties of activities.

Next, among all the entities present in a phrase, it is necessary to decide which is the 'focal concept'—the idea in the subject which is to be regarded as its key, the direct concern of the author. Other concepts are regarded as non-focal 'differences'. So in examples (b) and (c) above, the entities Rivers and Roses may be considered to be focal concepts. In example (a) the focus of attention is 'the spokes of bicycle wheels'. This must be further analysed by means of the 'possessive' or 'thing-to-part' relation into:

system—bicycles
sub-system—wheels
sub-sub-system—spokes

and its components assembled in that sequence. The remaining term in this example—Plating—is an attribute, and the complete heading is:

Bicycles. Wheels. Spokes. Plating

In example (b) there is a second entity, 'power station feedwater', analysed by the possessive relation into Power Station. Feedwater. This is linked to Rivers via the attribute term 'Contamination', which is an 'effect produced on' Rivers by the Feedwater. So the complete heading is:

Rivers. Contamination. Power Stations. Feedwater

In example (c) there are two other entities, DDT and Aphids, and two attribute terms. Predation links Roses and Aphids in an 'effect produced on . . . by . . .' relation; Control links Aphids and DDT in the same way. The full heading is thus:

Roses. Predation. Aphids. Control. DDT

The net effect of this detailed categorical analysis is to form index headings grouped by the principal systems of focal entities, with subgrouping determined by relational analysis—once again, in a way comparable to facet analysis. In all these principles of alphabetical indexing developed by Coates, Farradane and Austin, limited classificatory effect is achieved by methods similar to—and in fact derived from—those used to obtain more extensive grouping in faceted classification. The limitation is that, within any grouping, the sequence is alphabetical. No hierarchical sequence of terms is introduced.

CROSS-REFERENCES

The compound headings of the index are filed in alphabetical order. For any compound heading, immediate access is provided only from the first word. If the index user knows the noun order adopted, he can locate the particular compound he seeks. However, access from the secondary nouns is desirable for two reasons—to aid the less knowledgeable searcher, and to facilitate generic survey. This secondary access may be provided in the alphabetical index by cross-reference, as in the following examples:

(1) Antibiotics—Production—by Soil organisms
Cross-reference:
Soil organisms, *see also* Antibiotics—Production
(2) Plants—Disease—Control—by Antibiotics
Cross-references:
Antibiotics, *see also* Plants—Disease—Control
Disease control, *see also* Plants
(3) Pea seedling—Growth—effect of GA
Cross-references:
GA, *see also* Pea seedling—Growth
Growth, *see also* Pea seedling
Seedling, *see also* Pea
(4) Spectra—Atoms—Helium—Fine structure
Cross-references:
Fine structure, *see also* Spectra—Atoms—Helium
Helium, *see also* Spectra—Atoms
Atoms, *see also* Spectra

Instead of this 'chain indexing', Bernier suggested the systematic use of 'rotated entry', as follows:

Main entry:	Spectra—Atoms—Helium—Fine structure
Added entries:	Atoms—Helium—Fine structure—Spectra
	Helium—Fine structure—Spectra—Atoms
	Fine structure—Spectra—Atoms—Helium

By either of these methods, example (4) can be found no matter which term of the compound is searched.

Cross-reference is also used to facilitate generic survey. Cutter advised the regular addition of *see also* references from a term to each term subordinate to it, e.g.:

Building material, *see also* Wood, Brick, Stone, Cement
Stone, *see also* Granite, Sandstone, Marble

He advised only occasional reference in the reverse direction, from a term to a more generic term (e.g. Granite, *see also* Stone; Stone, *see also* Building material). However, others have recommended regular reference in both directions:

Building material, *see also* Wood
Wood, × × Building material.

Kaiser gave for each term 'higher collectives, lower specifics, and collaterals':

Stone<Building material
>Granite, Sandstone, Marble
=Wood, Brick, Cement

General *see also* references from a term to terms subordinate to it are also used, as in 'Birds, *see also* names of different birds, e.g. Plovers'.

Cross-references have come to their full flower in thesauri, which tend to replace the traditional subject heading lists. Whereas the latter presented meaningful groupings simply as 'see also' references, the former use a more refined relational analysis. A cross-referenced term may be part of the same classificatory hierarchy as the term referring to it—either the genus to which it belongs (broader term, BT) or a species of it (narrower term, NT). On the other hand, the two terms may be collaterally (species in the same genus) or semantically linked in some non-hierarchical way, and both these types are referred to as 'related terms' (RT). A typical thesaurus extract is as below:

Chemical properties	
BT	Chemistry
NT	Acidity
	Alkalinity
	Calorific value
	Chemical reactivity
	Heat of absorption
	Heat of combustion
	Heat of crystallisation
	Heat of formation
	Heat of fusion
	Heat of hydration
	Heat of reaction
	Heat of solution
	Heat of vaporisation

	Latent heat
	Thermochemical properties
	Valence
RT	Chemical bonds
	Molecular weight
	Optical activity
	Physico-chemical properties
	Thermodynamic properties

The result of this relational analysis is that the terms referred to as BT and NT all fall within the same conceptual facet as the referring term, whereas the RT may be either collateral terms in the same facet or terms from another facet: in effect, facet structure has partly been introduced into the displayed grouping. In some thesauri the 'concealed classification' lying behind the cross-reference structure is made explicit by a second sequence of terms which presents the hierarchy within each facet used, for example:

Chemical properties
- Acidity
- Alkalinity
- Chemical reactivity
- Thermochemical properties
  - Heat of absorption
  - Heat of crystallisation
  - Heat of reaction
    - Calorific value
    - Heat of combustion
    - Heat of formation
      - Heat of hydration
  - Heat of solution
  - Latent heat
    - Heat of fusion
    - Heat of vaporisation
- Valence

The explicit display of structure is taken one step further in the *Thesaurofacet* described by Jean Aitchison. The basic hierarchical relations of each index term are displayed in the schedules of a faceted classification. Thus the entry, 'Television camera tubes, MCE', in the alphabetical thesaurus refers to class MCE in the classified schedule:

M	Electronic engineering (BT)
MA	Electron tubes (BT)
MBT	Electron beam deflection tubes (BT)
MBV	Indicator tubes
MBW	Trochotrons
MC	Cathode ray tubes (BT)
MC2	Image converter tubes (RT)

MC4	Image intensifiers (RT)
MC6	Storage tubes (RT)
MCE	Television camera tubes
MCI	Colour tubes (NT)
MCL	Television picture tubes (RT)
MCO	Colour tubes (RT)
MCQ	X-ray tubes

The terms marked BT, NT, RT might be so referred in a typical thesaurus. In the *Thesaurofacet*, references to these broader, narrower and collateral terms in the same facet are *not* made in the alphabetical sequence. The latter is reserved for semantic relations to terms in *other* facets, thus:

Television camera tubes, MCE

RT	Photomultipliers
	Phototubes
	Television cameras
BT	(A) Television apparatus

The BT (A=auxiliary or additional) does not lie in the hierarchy to which Television camera tubes belongs, although these tubes are indeed a species of television apparatus. The net effect of the *Thesaurofacet* is to make the classificatory structure behind cross-references even more explicit.

In this discussion of the alphabetical subject index three forms of control have been described: (1) control of the type of heading used, with preference for the noun combination whenever possible; (2) control of the noun order by general rules which may be based on facet analysis; and (3) control of the cross-reference structure by a classification schedule. The fourth form of control is of the terms themselves—the choice of terms currently in use among those who will use the index, *see* references from synonyms, the combination of plural and singular forms (Grasses, Grass; Fungus, Fungi), and the use of bracketed scope words to differentiate homonyms such as Pitch (substance), Pitch (motion), Pitch (acoustics) and Pitch (angle). Given such controls, the alphabetical subject index is an invaluable tool for specific reference. Facilities for generic survey, provided by cross-reference, are secondary.

THE USE OF CLASSIFICATION IN ALPHABETICAL INDEXING

The controls suggested above amount to the basing of an alphabetical index on a faceted classification. I now want to illustrate this further by showing how facet analysis may reveal faults in alphabetical indexing and help the indexer to overcome them.

The alphabetical index to the journal *Chemical Abstracts* is perhaps the finest indexing tool available in any field of science, yet it is still possible to show that it might be improved. As an example, I present in list I the results of a search in one index for references to four subjects dealing with solvents: (1) solvents to remove grease from aluminium, (2) solvents for oils, (3) solvents for nitrocellulose, and (4) solvents for the separation of hydro-carbons.

List I: Subject headings in Chemical Abstracts

Solvents
  for A-substances, *1131*
  A-substances, *5003*
  Analysis of, 1367
  Critical solution temps, 663
  Degreasing, recovery, *5368*
  Diseases from, *2828*
  Effect of, on bactericidal action of phenol, 3566
    ,, on reaction of A with H, 4593
    ,, ,, spectra of dyes, 866
  Flash-point, effect of C, 4487
  for Hydrocarbon separation, 307, 5449
  for Ink, recovery of, in printing, 5093
  Manufacture of, *136*
  Molecular association of, 3474
  for Nitrocellulose, 2891
  Recovery, *3859*
  Removal from E, 3468

see also
Cleaning comps
Extraction
Lacquers
Petroleum refining
etc.

Cleaning compositions, *see also* Grease removal, etc.
  for Aluminium, *2663*
Grease removal
  from Aluminium, *2724*, *4584*
Extraction, *see also* Oils, etc.
Oils
  Solvents for, *1501*, *2079*
Lacquers, *see also* Nitrocellulose, etc.
  Nitrocellulose
    Solvent for, 2891
Nitrocellulose
  Solvent for, *1843*, 2891
Petroleum refining
  with Solvents, for separation of hydrocarbons, *2436*

Under solvents we find many references to solvents *for* various substances (those beginning with A, called A-substances in the list, and so on down the alphabet), and also to various substances (called A-substances and so on) which are *used* as solvents. Among these we find references of interest in our search (reference numbers also given):

for Hydrocarbon separation, 307, 5449
for Nitrocellulose, 2891

We also see other headings, such as Analysis and Critical solution. Among these, one looks promising:

Degreasing, recovery, 5368

At the head of the section on Solvents we are further directed to *see also* other headings, including those shown on the list. We follow these up.

Under Cleaning compositions we find a reference to Aluminium, 2663, and a direction to *see also* Grease removal. Here we find two more references to Aluminium, 2724 and 4584. The index does not direct us further forward, but if we look up the main heading Aluminium we find the following:

Cleaning, 2663, *4830*, *4838*, *5246*
Degreasing, 2724, *3607*, 4584

The four italicised references are new. Thus, of the eight references found, if we start from either Solvents or Aluminium the index itself only leads to half the number. Moreover, references 2724 and 4584 are reached from Solvents only by following two *see also* directions.

If we follow up the direction to *see also* Extraction, we find nothing of interest except a direction forward to Oils. Here we find two references to Solvents, 1501 and 2079. Lacquers provides one reference to a Nitrocellulose solvent, 2891, and leads to another, 1843. Petroleum refining gives one further reference, 2436.

The first feature of this search is the well-known scatter of related subjects in an alphabetical index, which necessitates an elaborate pattern of *see also* references. The second feature is that the indexing of compound subjects which contain more than one term is incomplete. None of the italicised references in list I is given under an alternative heading: thus *5368* is not repeated at Degreasing, nor *2828* at Diseases. In fact, of the 26 reference numbers in the list, 13 are not multiply indexed. We have already seen a similar proportion in the references to Solvents and Aluminium. The implication is that the provision of multiple approaches to compound subjects is incomplete and inconsistent.

In fact, this tiny sample has served to illustrate the main problems of subject indexing. Subjects are usually compounds of simpler terms—the four subjects of our search have, respectively, 4, 2, 2 and 3 terms. An index has two main jobs to do: to provide access to a reference from each term in the compound subject, and to link each term to related terms. The example given shows how even a very fine alphabetical index may fail to do the first job satisfactorily, and only does the second in the form of a ramification of *see also* references.

INDEXES BASED ON A FACETED CLASSIFICATION

Exactly the same 26 references may be indexed by means of a classification which has been built up by facet analysis. List II is in two parts. IIA gives index headings which would appear in a classified catalogue: class symbols, the verbal headings, and the reference numbers. IIB gives the alphabetical index by which the enquirer is led into the catalogue.

List IIA: Faceted classification: classified headings

E : 22a	Solvents
E : 22a : 11	,, :preparation, 136
E : 22a : 15	,, association, 3474
E : 22a : 3	,, analysis, 1367
E : 22a : 898	,, recovery, 3859
E : 22a ; c213/EC	,, flash point : effect of C, 4487
E : 22a ; c59	,, critical solution temperature, 663
E : 22aA	,, A-substance, 5003
E58 : c3/22a	Dyes : spectra : effect of solvent, 866
E61 : 6a	Hydrocarbons : separation agent (solvent), 307,2436,5449
E94 : 22a	Oils : solvent, 1501, 2079
E947 : 22a : 898	Grease : solvent : recovery, 5368
EA : 22a	A-substance : solvent, 1131
EA-H : 21/22a	A with H : reaction : effect of solvent, 4593
EAL : 899a	Aluminium : cleaning grease agent, 2663, 2724, 4584
EE : 22a : 6	E-substance : solvent : separation, 3468
EN : 22a	Nitrocellulose : solvent, 2891, 1843
L : 4(E : 22a)	Disease agent : solvent, 2828
L : 424 : 63P/E : 22a	Bacteria : killing by phenol : effect of solvent, 3566
M143 ; 591 : 22a : 898	Printing : ink : solvent : recovery, 5093

Facets used in List IIA

Substances, etc.

E58	Dyes
E61	Hydrocarbons
E94	Oils
E947	Grease
EA, etc.	
M143 : 591	Printing ink
L : 424	Bacteria causing disease

Operations and Actions

E : 11	Preparation
E : 15	Association
E : 21	Reaction
E : 22	Solution
E : 3	Analysis
E : 6	Separation
E : 898	Recovery
E : 899	Cleaning
L : 63	Chemotherapeutic killing

Properties

c213	Flash point
c3	Spectra
c59	Critical solution temp

Other symbols

/	Effect of
a	Agent

List IIB : Faceted classification : chain index

A-substance, EA
,, : solvent, E : 22aA
Aluminium, EAL
Analysis : solvents, E : 22a : 3
Association : solvents, E : 22a : 15
Bacteria : disease, L : 424
Bactericides : disease, L : 424 : 63
C-substance : effect on flash-point : solvents, E : 22a : c213/EC
Cleaning : aluminium, EAL : 899
Critical solution temperature : solvents, E : 22a ; c59
Disease, L : 4
Dyes, E58
E-substance, EE
Flash-point : solvents, E : 22a ; c213
Grease, E947
,, : cleaning : aluminium, EAL : 899
H-substance with A-substance, EA-H
Hydrocarbons, E61
Ink : printing, M143 ; 591
Nitrocellulose, EN
Oils, E94
Phenol : bactericide : disease, L : 424 : 63P
Printing, M143
Reaction : A with H, EA-H : 21
Recovery : solvents, E : 22a : 898
,, ,, : grease, E947 : 22a : 898
,, ,, : ink : printing, M143 ; 591 : 22a : 898
Separation : hydrocarbons, E61 : 6
Solvents, E : 22a
,, , effect on phenol : bactericide : disease, L : 424 : 63P/E : 22a
,, , ,, ,, reaction : A with H, EA-H : 21/22a
,, , ,, ,, spectra : dyes, E58 ; c3/22a
,, for A-substance, EA : 22a
,, ,, E-substance, EE : 22a
,, ,, grease, E : 947 : 22a
,, ,, ink : printing, M143 ; 591 : 22a
,, ,, Nitrocellulose, EN : 22a
,, : disease, L : 4(E : 22a)
,, : separation : hydrocarbons, E61 : 6a
Spectra : dyes, E58 ; c3

We note first that the classified index avoids the scattering of related references found in the *Chemical Abstracts* index. For example, the references to 'solvents for the separation of hydrocarbons' are all in class E61 : 6a. Second, the index to the classified catalogue enables us to reach a reference no matter what term of the subject we start from, as follows:

Solvents : separation : hydrocarbons, E61 : 6a
Separation : hydrocarbons, E61 : 6
Hydrocarbons, E61

Each term in the chain is indexed. If we enter the index by the last term ('Hydrocarbons'), we are directed to E61 in the catalogue. Once there, we follow its sub-divisions till we come to E61 : 6a, with the references we need. This is the result of alphabetically indexing the classified catalogue by chain procedure.

A faceted classification and its chain index therefore provide a consistent solution of the problems mentioned earlier: access to a reference is provided from each term in a compound subject, and terms which are related generically are brought together in the catalogue. When the same term appears in more than one genus, all its locations in the catalogue are brought together in the alphabetical index.

For specific searches, a classified catalogue is indirect, and the advantages of bringing together generically related terms are outweighed by the disadvantages of always having to pass from the alphabetical index to the classified catalogue before reaching the references.

To eliminate the classified catalogue, we can either sacrifice the generic relations; or we can indicate them with *see also* references from a specific term to its immediate genus (e.g. Grease *see also* Oils); or we can transform the combination of lists IIA and IIB into a single alphabetical catalogue, list III. This is an alphabetical subject index comparable to that of *Chemical Abstracts*, but the entries have now been made consistently, and access to a reference is provided from every term in a compound subject.

List III: Alphabetical subject headings based on lists IIA–B

A-substance : solvent for, 1131
A-substance as solvent, 5003
A-substance with H-substance : reaction : effect of solvent, 4593
Aluminium : cleaning grease agent, 2663, 2724, 4584
Analysis : solvents, 1367
Association of solvents, 3474
Bacteria : killing by phenol : effect of solvent, 3566
Bactericide : phenol : effect of solvent, 3566
C-substance : effect on flash-point of solvents, 4487
Cleaning aluminium, 2663, 2724, 4584
Critical solution temperature of solvents, 663
Disease caused by solvent, 2828
Dyes : spectra : effect of solvent, 866
E-substance, solvent for : separation of, 3468
Flash-point of solvents : effect of C-substance, 4487
Grease, solvent for : recovery, 5386
Grease : cleaning off aluminium, 2663, 2724, 4584
H-substance with A-substance : reaction : effect of solvent, 4593
Hydrocarbons : separation solvent, 2436, 307, 5449
Ink, printing, solvent for : recovery, 5093
Nitrocellulose, solvent for, 2891, 1843

Oils, solvent for, 1501, 2079
Phenol as bactericide : effect of solvent, 3566
Printing ink, solvent for : recovery, 5093
Reaction, A with H : effect of solvent, 4593
Recovery of solvents, 3859
,, ,, ,, for grease, 5368
,, ,, ,, ,, ink, printing, 5093
Separation of hydrocarbons : solvent, 2436, 307, 5449
Solvents : A-substance, 5003
,, : effect on phenol as bactericide, 3566
,, ,, ,, reaction, A with H, 4593
,, ,, ,, spectra of dyes, 866
,, for A-substance, 1131
,, ,, E-substance : separation, 3468
,, ,, grease : recovery, 5368
,, ,, ink, printing : recovery, 5093
,, ,, nitrocellulose, 2891, 1843
,, producing disease, 2828
,, : separation of hydrocarbons, 2436, 307, 5449
Spectra : dyes : effect of solvent, 866

THE CLASSIFIED INDEX

Indexing a document by means of a faceted classification may be illustrated by using the Soil Science schedules in Appendix B to index the subject, 'changes in pH of a podzolic soil as influenced by superphosphate dressings'.

(1) The first step is to set out the subject in terms of the categories and relations used in the classification, thus:

change:	general property
pH:	property
podzol:	kind of soil
influenced by:	relation
superphosphate:	material
dressings:	operation

(2) Then the categories are rearranged into the combination order adopted in the classification, the terminology used in the schedules is substituted, and the notational symbols for each term are recorded, thus:

Category	*Term in subject*	*Term in scheme*	*Symbol*
kind of soil	podzol	podzol	9s
property	pH	pH	6r
general property	change	variation	1f
relation	influenced by	effect of	/
operation	dressing	amendment	4m
material	superphosphate	superphosphate	4nm

(3) The possibility of expressing the subject more succinctly is then considered. 'The effect of superphosphate on pH' adequately expresses 'the influence of superphosphate on changes in pH', so the term 'variation' can be dropped. 'Superphosphate' is already listed in the schedules as an amendment material, so 'amendment' can be dropped.

(4) The remaining symbols are then synthesised into a class number, thus: 9s6r/4nm.

Once familiarity with a classification scheme is achieved, the formality of this approach is dropped, and class numbers are synthesised much more directly. But in all difficult cases a systematic analysis of this kind is helpful. Complex examples are given by Ranganathan in his *Prolegomena* and elsewhere.

INDEXING ORDER AND COMBINATION ORDER

The following section of a classified index illustrates the indexing order.

Soils by texture	9tm
Loamy sand	9v
effect of clay	9v/y
drainage	9v4d
effect of porosity	9v4d/6i
by mole	9v4db
erosion	9v5u
effect of moisture	9v5u/6p
by water	9v5ud
rate	9v5ud1d
exchange capacity	9v6q
effect of nitrogen	9v6q/7f
effect of pH	9v6q/r
salinity measurement	9v6s3b
electrical	9v6s3bE
humus	9v7i
effect of irrigation	9v7i/4f
analysis	9v7i3m
adsorption by granules	9v7i5n8mb
pH	9v7i6r
crumb	9v8m
formation	9v8m5b
effect of humus	9v8m5b/7i
moisture content	9v8m6p
nutrient	9v8m7d
measurement	9v8m7d3b
availability	9v8m7db
variation	9v8m7dble
fine sand particles	9v8m–rd

desert	9v–mb
survey	9v–mb4b
humification	ov–mb5h
porosity	9v–mb6i
fine sand particles, trace elements	9v–mb8rd7e
produced by rock weathering	9v(7m5c)
Sandy loam	9vn

The order of small letters after the same numeral—e.g. q and s after 9v6—reflects the order of terms in the facet: here the property of 'exchange capacity' precedes that of 'salinity'. The order of the numerals, which are facet indicators, determines that the subjects subsidiary to 'loamy sand' will be taken in a certain order:

4 operations such as 'drainage'
5 processes such as 'erosion'
6 properties such as 'salinity'
7 constituents such as 'humus'
8 structures such as 'crumb'

The ordinal values of the phase relation symbol (stroke), of the inner-facet compound symbol (hyphen) and of the subject device (bracket) are assigned so that compounds with the first of these ('effect of clay') precede the other subsidiary subjects, and compounds formed by the other two devices ('desert loamy sand' and 'loamy sand produced by rock weathering') follow the others.

The combination order of facets in a compound has a profound effect on indexing order. Here it determines that all the effects, operations, processes, properties, constituents and structures specific to 'loamy sand' will be grouped together in what is hoped to be 'a useful togetherness'. If facet 9, Kinds of soil, were cited *after* the others, then a given operation, say 'drainage', would be subdivided by all the kinds of soil drained, but with the combination order adopted 4d, Drainage, only becomes the primary facet if the soil facet is vacant.

Though all references to 'loamy sand' are grouped together, references to 'drainage' will be found both at 4d, Drainage, when the soil facet is vacant, and under every kind of soil drained. A term appearing still later in the combination order (e.g. 'pH', cited third at 9v7i6r and at 9v6q/r) will be distributed even more widely. Such terms will recur at widely separated parts of the classified index—thus in the case of 'Amendment':

Amendment	4m
pH	6r

Podsol	9s
pH	9s6r
effect of amendment	9s6r/4m
Loamy sand	9v
exchange capacity	9v6q
effect of pH	9v6q/r
humus	9v7i
pH	9v7i6r
Sandy loam	9vn
amendment	9vn4m

Such a classified index permits of specific reference: at each specific entry references to that subject are entered. It only permits of some generic survey: for example, everything related to 'nutrients in loamy sand crumbs' is gathered below 9v8m7d; everything on 'loamy sand crumbs' below 9v8m; everything on 'loamy sand' below 9v; everything on 'soils classified by texture' below 9tm; and everything on named soils below 9. For all these subjects, the classified index satisfies the second kind of generic survey—reference to 9v8m7d from terms which include it along the chain 9v8m7d→9v8m→9v→9tm→9. But the classified index alone does not satisfy this kind of generic survey for reference along such chains as 9vn4m→4m→4. Not everything on 'Amendment' is gathered below 4m. Such terms are scattered.

This problem of 'distributed facets' is solvable in two ways. The first solution is to make multiple entries in the classified index, which must contain not only 9vn4m but also 4m9vn. A quaternary compound such as 9v7i5n8mb must be entered at least three more times: if every permutation is to be entered, 23 more times. In the second solution only one entry is made in the classified index, and cross-reference entries are made in a complementary alphabetical chain index. This procedure will be described a little later.

Before coming to it, we may consider whether the indexing order of facets should be related to their combination order in a compound. Ranganathan lays it down as a matter of principle that the combination order of facets should be the reverse of the preferred indexing order. He regards the series *time*, *space*, *energy*, *matter* and *personality* as one of increasing 'concreteness', and believes that this indexing order is the most helpful. Therefore compound subjects should be listed in the following order:

(1) Energy: Space: Time
(2) Matter: Space: Time
(3) Matter: Energy: Space: Time
(4) Personality: Space: Time
(5) Personality: Energy: Space: Time
(6) Personality: Matter: Energy: Space: Time

This indexing order is automatically ensured if (a) the combination order of categories is the reverse of preferred indexing order—i.e. as in subject (6), and (b) the facet indicators for the five categories increase in ordinal value from *time* to *personality*.

Here I suggest a different approach. Combination order should be decided independently, in ways already discussed in Chapter 2, so that we have an established facet formula, and the preferred indexing order need not necessarily be related to combination order. However, it is possible to argue that (a) the later a category in the facet formula, the more 'general' it is likely to be, and (b) the more 'general' a subject is, the earlier it should appear in indexing order. If this argument is correct, we can adopt Ranganathan's convention and say that the ordinal values of facet indicators should decrease as we start from *substance* and proceed along the facet formula. The Soil Science schedules adopt this convention, as shown:

Combination order	1	2	3	4	5
Facet	Soil	Structure	Constituents	Properties	Processes
Indicator	9	8	7	6	5

Combination order	6	7	8
Facet	Operations	Techniques	General
Indicator	4	3	1

On the other hand, the simplest convention of all to adopt is that combination order and indexing order are identical. The facet indicators are then numbered as in the top row of the table.

CHAIN INDEXING

To deal with the problem of distributed terms, Ranganathan has developed the technique of constructing—as a complement to the classified index—an alphabetical index systematically compiled by chain procedure. An example will make the method clear.

The heading Soft palate—Cancer—Radium treatment—Statistical study is represented in the Colon classification by the class number L2153 : 4725 : 63129 : B28, which is built up as follows:

L	Medicine
L2	Digestive system
L21	Mouth
L215	Palate
L2153	Soft palate
L2153 : 4	Disease
L2153 : 47	Structural
L2153 : 472	Tumour

L2153 : 4725	Cancer
L2153 : 4725 : 6	Treatment
L2153 : 4725 : 63	Chemical
L2153 : 4725 : 631	Element
L2153 : 4725 : 6312	Group II
L2153 : 4725 : 63129	Radium
L2153 : 4725 : 63129 0a B	Mathematical study
L2153 : 4725 : 63129 0a B2	Algebra
L2153 : 4725 : 63129 0a B28	Statistical

By chain procedure, an alphabetical index heading is provided for each link in the chain of terms. In each case the heading is followed by a reference to the appropriate number in the classified index.

(1) Algebraical study : radium treatment : cancer : soft palate—see L2153 : 4725 : 63129 0a B2
(2) Cancer : soft palate—see L2153 : 4725
(3) Chemical treatment : cancer : soft palate—see L2153 : 4725 : 63
(4) Digestive system—see L2
(5) Disease : soft palate—see L2153 : 4
(6) Elements : treatment : cancer : soft palate—see L2153 : 4725 : 631
(7) Group II elements : treatment : cancer : soft palate—see L2153 : 4725 : 6312
(8) Mathematical study : radium treatment : cancer : soft palate—see L2153 : 4725 : 63129 0a B
(9) Medicine—see L
(10) Mouth—see L21
(11) Palate—see L215
(12) Radium treatment : cancer : soft palate—see L2153 : 4725 : 63129
(13) Soft palate—see L2153
(14) Statistical study : radium treatment : cancer : soft palate—see L2153 : 4725 : 63129 0a B28
(15) Structural disease : soft palate—see L2153 : 47
(16) Treatment : cancer : soft palate—see L2153 : 4725 : 6
(17) Tumour : soft palate—see L2153 : 472

This procedure does not use all the permutations of terms possible, but it is claimed that if, for example, the searcher has formulated his subject as Statistical study—Cancer—Soft palate—Radium Treatment, the index entry (14) will catch his eye, and he will be directed to the correct place in the classified catalogue despite having used an incorrect combination order. If the searcher has left out certain terms in formulating his subject and consults the index at Cancer—Soft palate, entry (2) will direct him to L2153 : 4725. At this point in the classified catalogue the further sub-divisions of this term will catch his eye and lead him down the chain to the exact subject of his search. Finally, if he has used a broader heading to approach his subject, such as Palate or Disease, entries (11) or (5) will land the searcher in the right chain.

By means of chain procedure, a solution to the problem of completely flexible generic survey is offered. Each term in a compound subject becomes an entry point in the alphabetical index, and at this point references to all the places in the classified index where it appears are gathered. For example, in the chain index for Soil Science would appear the following:

Amendment—see 4m
  effect on : pH : podsol—see 9s6r/4m
  sandy loam—see 9vn4m
pH—see 6r
  effect on : exchange capacity : loamy sand—see 9v6q/r
  humus : loamy sand—see 9v7i6r
  podsol—see 9s6r

Our requirement that there be reference to a compound ABCD from any of its component terms is met, and there is moreover generic reference along chains such as 9vn4m→4m.

The relative economy of chain procedure should also be noted. Multiple entry in the classified index under each term of a compound subject, ABCD requires *for each subject* as many entries as there are terms in a compound. Chain indexing requires for a compound ABCD four alphabetical entries, for DCBA, CBA, BA and A. But these entries also serve for subjects ABC, AB and A. If a classification has been designed with close attention to literary warrant, we may assume that as the index grows there will appear a reference to every possible subject—i.e. there will be a reference under each of the subjects A, AB, ABC and ABCD. These four subjects need only four alphabetical and four classified entries. Multiple entry in the classified index would need ten entries. In general terms, chain procedure for compounds of up to N terms tends to need $2N$ entries, multiple entry to need $1+2+\ldots+N$ entries, which is greater than $2N$ if $N>3$.

Despite these virtues, there are difficulties with the use of chain indexes, recently discussed by Sweeney and by Mineur.

BIBLIOGRAPHY

J. AITCHISON. 'The thesaurofacet—a multipurpose retrieval language tool'. *J. Docum.*, 1970, **26,** 187–202

D. W. AUSTIN and P. BUTCHER. *PRECIS—A Rotated Subject Index.* London, 1969. The PRECIS system has been further developed by AUSTIN, as described in *J. Docum.*, 1974, **30**, 47–102

E. J. COATES. *Subject Catalogues—Headings and Structure.* London, 1960

J. E. L. FARRADANE. 'Concept organisation for information retrieval'. *Inf. Stor. Retr.*, 1967, **3,** 297–314

J. METCALFE. *Information Indexing and Subject Cataloguing.* New York, 1957

B. W. MINEUR. 'Relations in chains'. *J. Lib.*, 1973, **5,** 175–202

R. K. OLDING (ed.) *Readings in Library Cataloguing.* London, 1966. (Includes selections from Cutter, Kaiser, Dewey, Hulme, Bliss and Ranganathan)

M. PREVOST. 'An approach to theory and method in general subject heading'. *Lib. Qrly.*, 1946, **16,** 140–51

R. SWEENEY. 'Indexing the classified catalogue'. *Cat. and Index*, 1970, No. 19, 10–13

6

CLASSIFICATION IN POST-COORDINATE SYSTEMS

The indexes discussed in Chapter 5 have all consisted of a series of subject statements, each topic being represented by a string of coordinated terms, e.g.

Spectra—Atoms—Helium—Fine structure
Acid/: Hydrolysis/- Sucrose/: Glucose
Solvents, for Ink, recovery of, in printing
9v–mb8rd7e

As has been pointed out, search access to each topic must be provided from the secondary terms within it, and various cross-reference techniques have been described, such as rotated entry or chain indexing.

There are two problems arising from this. First, as the number of terms in a string grows, so the number of index entries per topic increases, even though chain indexing tries to minimise this. Second, for reasons of economy, none of the cross-reference techniques attempts to provide an entry for every possible permutation of terms in the string. For example, rotation of a string ABCDE would produce only the extra entries BCDEA, CDEAB, DEABC, EABCD. Suppose that a searcher was interested in a topic expressible by the coordination of A and C. Whether he enters the index at A or C, he may overlook entries that do include both terms, but in which they are not adjacent to each other in the string. This is not just a theoretical hazard—practical experience of the use of indexes consisting of such 'pre-coordinated' strings has confirmed it.

THE POST-COORDINATE APPROACH

For this and other reasons, an alternative method of indexing has developed over the past few decades, the post-coordinate (or, more briefly, the coordinate) approach. Instead of constructing a pre-coordinate entry such as

Solvents, for ink, recovery of, in printing, 5093

and adding either extra entries such as

Ink, printing, solvent for, recovery, 5093

or *see* references, a series of unit term entries is prepared:

Solvents, 5093
Ink, 5093
Recovery, 5093
Printing, 5093

When a search is made for 'recovery of solvents for printing ink', the four unit entries are examined and matched, to reveal that they carry at least one document number in common, 5093. The same set of entries permits 'post-coordinate' search for any combination of the unit terms, thus avoiding both the uneconomic multiplication of entries and the search hazards encountered in pre-coordinate indexes.

Since the terms representing a topic are no longer brought together to form a compound heading, the direct problems of term sequence—discussed as the combination order of categories in Chapter 2, and as noun order in Chapter 5—are no longer encountered. Simple post-coordination of the unit terms 'Dyestuffs', 'Bacteria', 'Destruction', cannot distinguish between the topics 'destruction of dyestuffs by bacteria' and 'destruction of bacteria by dyestuffs', and the introduction of devices to make this distinction once again raises the question of relations between categories. This will be discussed later.

The terms chosen for use in a post-coordinate system—whether they are written on cards, or coded into punchcards or magnetic tape—can, just as in any other type of index, be drawn either direct from natural-language text or from a controlled vocabulary, alphabetical or classified. Cross-reference relationships between terms remain equally important in search, whether the terms themselves are coordinated at the indexing or at the retrieval stage. Indeed, as noted in Chapter 5, it is the thesaurus—which has developed concurrently with the post-coordinate approach—that has considerably increased the number of relations displayed in an alphabetical vocabulary.

This is true even in systems where the terms recorded at input are simply taken from natural language text. For example, a post-coordinate system may be based upon the words occurring in the titles of articles. A search may be required for articles relating to

Mycology. This will necessitate the identification of many words with a mycological meaning, that might be expected in titles—for example, 'Fungus', 'Spore', 'Mycelium', 'Yeast', 'Mould', 'Blight', 'Mildew', 'Mycosis', 'Penicillium', and many more. The list may be constructed partly from the searcher's knowledge, but he will be aided by a thesaurus that displays as many semantic relations as possible, from which he may select those appropriate to his interests.

TYPES OF THESAURUS

The general structure of an alphabetical relational thesaurus has already been considered in Chapter 5 (pp. 108–109). Often there are two sequences—an alphabetical list of terms showing BT, NT and RT relations, and a hierarchical list. A highly developed example of this, the *Thesaurofacet*, has already been described. Facet analysis to develop a thesaurus has also been used in computer-based systems—for example, for the petroleum industry. Both the subject authority list of the American Petroleum Institute (described by Mulvihill and Brenner) and the *Exploration and Production Thesaurus* of the University of Tulsa rely on the same facet organisation. Martinez and Helander explain that all terms in each list are allocated to one of the following ten basic concepts (categories): Common Attribute, Earth and Space Concepts, Economic Factor, Equipment, Material, Operating Condition, Organism, Phenomenon, Process, and Property.

'Once these ten basic concepts had been defined, it was a simple matter to determine into which of these categories each individual descriptor should fall. Equipment consisted essentially of any tool, device or object used for any purpose whatsoever. Material, in contrast, was the basic substance of which any object consisted, or was made. For example, the term PIPELINE would be considered as Equipment, whereas STEEL, from which the pipeline was made, would be Material. A term such as WELL, although not normally thought of as Equipment, would be classified as such because it is constructed and serves a useful purpose. The facet Phenomenon was taken to include all naturally-occurring happenings or phenomena, as contrasted to Process which was designed to include any man-directed or adapted happening, usually to accomplish a specific objective. Thus, the term ABSORPTION would be used to characterise abstracts dealing with this particular phenomenon, its theory and perhaps laboratory studies thereof. The term ABSORPTION PROCESS would refer to abstracts dealing with commercial utilisation of absorption, for a specific purpose, such as the separation of components in a natural gas stream. Another pair of related facets consisted of Property and Common Attribute. In this case Property refers to an inherent characteristic of an object or material, such as the hardness or ductility of a steel rod. A Common Attribute, in contrast, is considered as an accidental or variable characteristic of an object or material, such as the length or thickness of a steel rod. The Operating Condition facet includes variable or controlled phenomena or properties, usually in conjunction with an industrial process, such as TEMPERATURE, PRESSURE, VELOCITY, EFFICIENCY, etc. The remaining three facets are self-explanatory. Economic Factor is used to categorise various fundamental concepts involved in economic considerations, such as COST, PROFITABILITY, SUPPLY and DEMAND, INSURANCE, LEGAL CONSIDERATION, RISK, etc. It should be noted that business operations such as ACCOUNTING, ADMINISTRATION, COMMUNICATION, RESEARCH, TRAINING, etc. are not considered as Economic Factors, but are properly placed in the Process facet. The Organism facet embraces all organisms, plant or animal, living or dead. This separate facet was set up to accommodate the large number of terms required to categorise fossiliferous remains described in geological articles abstracted by Petroleum Abstracts. Similarly, Earth and Space Concepts was set up as a facet to accommodate the many terms required to describe geological facets and structures.'

A comparable analysis by Mulvihill and Brenner is shown in *Table 6.1*, and a brief extract from the *process* facet shows it to be in alphabetico-classed sequence:

(Process)
  (Physical operation)
    Water scrubbing
    Water treating
      Desalination
      Water thickening
    Wedging
    Welding
      Automatic welding
      Electric arc welding
    Well completion
      Liner completion
        Screen liner completion
        Slotted liner completion

An example of the alphabetical listing in the *E and P Thesaurus* is shown in *Table 6.2*.

In a restricted field, for a punchcard system, a brief and simple faceted vocabulary may be used—as in the US Patent Office system for coding chemical tests (*Table 6.3*). An on-line search system developed within the British Steel Corporation uses an

Table 6.1 FACETS IN A.P.I. VOCABULARY

- [PROCESS]
- (Goal-oriented activity)
  - Industrial process
  - (a complete multistep operation)
    - e.g. Isoforming
  - Chemical reaction
  - (Unit process)
    - e.g. Isomerization
  - [Physical change]
  - (a unit operation)
    - Physical separation
      - e.g. Distillation
    - (Miscellaneous)
      - e.g. Size reduction
  - Analytical method
    - e.g. Diffraction analysis
  - Business operation
    - e.g. Advertising
  - [Miscellaneous]
    - e.g. Maintenance
- [PHENOMENON]
- (a natural occurrence, happening)
  - e.g. Corrosion
  - e.g. Electricity
  - e.g. Engine noise
- MATERIAL
  - [By use]
    - e.g. Motor fuel
  - [By composition]
    - e.g. Gasoline
    - e.g. Methanol
  - [By phase]
    - e.g. Liquid
    - e.g. Emulsion
- EQUIPMENT
  - e.g. Petroleum refinery
  - e.g. Heat exchanger
  - e.g. Tube
- LIVING ORGANISM
  - Microorganism
  - Plant
  - Animal
  - Man
- ECONOMIC FACTOR
  - e.g. Supply
  - e.g. Exports
  - e.g. Employment
- [PLACE]
- [PROPERTY]
- (of materials)
  - Physical property
    - e.g. Melting point
    - e.g. Taste
  - Composition
  - (of mixtures)
    - e.g. Sulfur content
    - e.g. Aromatic
  - [Structure]
    - Chemical structure
      - e.g. Aromaticity
    - Polymer structure
      - e.g. Crosslinked polymer
    - Microstructure
      - e.g. Grain structure
      - e.g. Hexagonal crystal
- OPERATING CONDITION
- (of processes and equipment)
  - e.g. Temperature 100 to 200 C
  - e.g. Velocity
- [COMMON ATTRIBUTE]
  - Shape
    - e.g. Elliptical
  - Arrangement
    - e.g. Parallel
  - Manufactured form
    - e.g. Pellet
  - Portion/Position
    - e.g. Top
  - Direction
    - e.g. Crosscurrent
  - Operation
    - e.g. Hydraulic
  - Size
    - e.g. Micro
  - Quantity
    - e.g. Two
  - Numerical relation
    - e.g. Equation
  - Time/Frequency
    - e.g. Simultaneous
  - Quality
    - e.g. Standard quality
  - [Origin]
    - e.g. Synthetic
  - [Market/Affiliation]
    - e.g. European Common Market
  - [Science/Technology/Industry]
    - e.g. Geologic
    - e.g. Petrochemical industry

alphabetico-classed thesaurus with up to six hierarchical levels—an excerpt is shown in *Table 6.4*. The thesaurus is held internally by the computer. If the term 'coke ovens' is keyed in for search, all the terms specific to it in the hierarchy are presented to the

Table 6.2 EXTRACT FROM E & P THESAURUS

```
(A)  SEISMIC REFLECTION METHOD
(B)     **  PERTAINING TO USE OF SEISMIC
              REFLECTION. FOR SEISMIC
              REFLECTION WAVE PROPAGATION
              USE (REFLECTION (SEISMIC)).
(C)     UF  REFLECTION METHOD
        UF  REFLECTION PROSPECTING
        UF  REFLECTION SHOOTING
        UF  VERTICAL REFLECTION SURVEY
         WITH  VERTICAL
(D)     NT  ECHO SOUNDING
(E)     BT  SEISMIC EXPLORATION
              GEOPHYSICAL EXPLORATION
               EXPLORATION
        SA  CONTINUOUS PROFILING
        SA  DIP SHOOTING
        SA  HIGH FREQUENCY
        SA  MULTIPLE REFLECTION
        SA  REFLECTING BED
        SA  REFLECTION (SEISMIC)
        SA  REFLECTION RECORD
        SA  SEISMIC REFLECTION SURVEY
(F)     SA  SEISMIC REFRACTION METHOD
        SA  SEISMIC VELOCITY COMPUTATN
        SA  TRAVEL TIME ANOMALY
        SA  VELOCITY PROFILING

     SEISMIC REFLECTION PROSP
(G)     *   SEISMIC REFLECTION
              PROSPECTING
(H)     USE SEISMIC REFLECTION METHOD
```

Explanation

(A) Descriptor or term.

(B) ** -- This notation (scope note) identifies explanatory material, defining or limiting the meaning of a particular descriptor that might otherwise prove ambiguous.

(C) UF -- This notation (use for) indicates that the primary term, a valid term, is the preferred descriptor which should be used in place of the secondary descriptor bearing this notation.

(D) NT -- This notation (narrower term) indicates that the secondary term so designated is a more specific subdivision of a broader class, as represented by the primary term.

(E) BT -- This notation (broader term) indicates that the secondary descriptor so designated represents a more generalized class, of which the primary term is a member.

(F) SA -- This notation (see also) is used to designate related valid descriptors which are included primarily as supplementary leads for indexers or searchers, trying to locate a desired concept in the thesaurus.

(G) * -- This notation is used to designate the completely spelled-out version of an otherwise abbreviated descriptor.

(H) USE -- This indicates that the primary term is invalid, and that the secondary term listed should be substituted therefor.

Table 6.3 US PATENT OFFICE CHEMICAL TESTING VOCABULARY

U. S. DEPARTMENT OF COMMERCE
Patent Office

csg

C H E M I C A L   T E S T I N G   C O D I N G   S H E E T

D.P.	H			230.000 253.000	256.000	
REISSUE	E		Document Number	231.000 254.000	258.000	Accession Number
PATENT	A	US		023 232.000 255.000		C 4
Col.	1	23	4 5 6 7 8 9 10	11-3	14-9	21 22 23 24 25-6

MATL BY USE/FUNCTION/SOURCE

Col.	Code	Term
27	8	Adulterant
27	9	Amino-acid
28	Y	Biological
	YX	.Blood
	YX0	..Albumin
	YX1	..Hemoglobin
	YX2	..Rh or blood type
	YX3	..Sugar
	YX4	..Oth sp blood test
	Y5	.Breath
	Y6	.Gastric juice
	Y7	.Glandular
28	Y78	..Thyroid
29	X	.Urine + 28/Y
	X0	..Sugar + 28/Y
	1	.Oth biol + 28/Y
	2	Veg (non-food)
	23	.Cellulose, cork
	4	Bldg or road
	45	.Brick or rock
	46	.Cement or concrete
	47	.Glass or oth ceramic
	48	.Soil
29	49	.Other bldg
30	5	Catalyst
	6	Coating
	7	Enzyme
	8	Explosive
30	9	Fertilizer
31	0	Food
	01	.Beverage
31	02	.Dairy or eggs
	03	.Fish or meat
	04	.Vegetable
	05	.Other food
	6	Fuel
	7	Gas
	8	Ink, paint or dye
31	9	Leather
32	0	Lubricant
	1	Oil
	2	Process stream
	3	Resin
	4	Rubber
	5	Textile
	56	.Animal fiber
	7	Water
32	8	Other

CONDITION OR PROPERTY

Col.	Code	Term
32	9	Acidity
33	Y	Age dating
	X	Aroma or odor
	0	Corrosion
	1	Decay
	2	Defect, flaw, leak
	3	Fertility
	4	Geochem explor
	5	Interferent
	6	Lubricity
	7	Pregnancy
	8	Viscosity
33	9	Other

MATERIAL ANALYZED

Col.	Code	Term
34	Y	Inorganic
	YX	.Radioactive cpd
	Y0	.Inert gas
	Y1	.Rare earth cpd
	Y2	.B and cpds
	Y3	.P and cpds
	Y4	.Si and cpds
	Y5	.N and cpds
	Y6	.C and cpds
	Y7	.Hal and cpds
	Y8	.Se, Te and cpds
34	Y9	.S and cpds
35	7	.O and cpds + 34/Y
	8	.H and cpds + 34/Y
35	9	.Met + 34/Y
36	X	Organic
	X0	.Metal
	X1	.Boron
	X2	.Sulfur
	X3	.Silicon
	X4	.Nitrogen
	X45	..Aromatic
	X46	..Aliphatic
	X7	.Halogen
	X8	.Oxygen
36	X89	..C,H, O
37	2	.C, H +36/X
	23	..Unsat + 36/X
37	4	.Other nonmet + 36/X

ANALYZER BY USE

Col.	Code	Term
37	5	Color
	6	Catalyst
	7	Enzyme
	8	Indicator dye
37	9	Liquid crystal
38	2	Special form
	23	.Laminated
	4	Surfactant
	5	Tracer mat add
	56	.Radioactive
	57	.Phys detecable
	58	.Reactable
38	9	Other

ANALYZER (CHEM)

Col.	Code	Term
39	Y	Inorganic
	YX	.Radioactive
	Y0	.Inert gas
	Y1	.Rare earth
	Y2	.B and cpds
	Y3	.P and cpds
	Y4	.Si and cpds
	Y5	.N and cpds
	Y6	.C and cpds
	Y7	.Hal and cpds
	Y8	.Se,Te and cpds
39	Y9	.S and cpds
40	7	.O and cpds +39/Y
	8	.H and cpds +39/Y
40	9	.Met +39/Y
41	X	Organic
	X0	.Metal
	X1	.Boron
	X2	.Sulfur
	X3	.Silicon
	X4	.Nitrogen
	X45	..Aromatic
	X46	..Aliphatic
	X7	.Halogen
	X8	.Oxygen
41	X89	..C,H, O
42	1	.C,H + 41/X
	12	..Unsat + 41/X
42	3	.Other nonmet +41/X

BASIS FOR ANALYSIS

Col.	Code	Term
42	4	Sorption
	45	.Ion exchange
	46	.Paper or thin
	7	Calorimetry
	8	Chemilumin
42	9	Color reaction
43	0	Combustion
	01	.Plasma
	2	Comparison
	3	Condens nucl
	4	Displacement
	5	Evol gas
	6	Extraction
	7	Flame analysis
	8	Pptn coagn
43	9	Titration
44	Y	Phys meas
	YX	.Time var
	Y0	.Lgth or ar
	Y1	.Vol, wt or d
	Y12	..Appendage
	Y3	.Pressure
	Y4	.Temperature
	Y45	..Gradient
	Y6	.Thermal cond
	Y7	.Viscosity
44	Y8	.Flame ioniz
45	Y	.Optical (index also 44/Y)
	YX	..Color/absorb
	Y0	..Spectroscopy
	Y1	..Sensor
	Y12	...Photocell
	3	.Non-visible wave
	34	..Infra-red
	35	..Ultra-violet
	36	α/β/γ
	37	..Sensor (index also 44/Y)
	8	.Elec cond
45	9	.Other phys

MANIPULATION OR DEVICE THEREFORE

Col.	Code	Term
46	Y	Preparation
	YX	.Sepn or purif
	YX0	..Centrif or seb
	YX1	..Demist fog
	YX2	..Distill
	YX3	..Dry
	YX4	..Impurities
	YX5	..Other sepn
	Y6	.Phase change
	Y7	.Heat or cool
	Y8	.Digest or cook
46	Y9	.Irradiate
47	X	.Pressure
	0	.Chem react (index also 46/Y)
	01	..Catalytic
	02	..Neutralize
	3	.Control flow
	4	.In train
	5	.Titration
	6	.Aliquot
	7	.Permeable
	8	.Pump
47	9	.Carrier gas
48	0	.Holding vessel
	01	..Trans vessel (index also 46/Y)
	02	..Trans contents
	03	..Mix or stir
	04	..Empty
	05	..Rack
	056	...Trans racks
	7	Environmental
	8	Capillary
48	9	Det station
49	Y	Automated anal
	YX	.Train
	Y0	.Central
	Y01	..Multiple
	Y02	..Program
	Y03	.Decoder
	Y04	.Computer-assist
	Y5	.Process control
	Y56	..Chem process
	Y57	..Non-chem proc
		..Signal generated by:
	Y58	...T, P or concn
49	Y59	...Liquid level
50	1	...Flow rate (index also 49/Y5)
	2	...pH,cond
	3	...Summary
		..Signal fed:
	4	...Back
	5	...Forward
	6	...To detn pt takeoff
		..Signal affects:
	7	...Input feed
	8	...Other conditions
50	9	..Computer-assist
51	5	.Monitoring + 49/Y
	6	.Other use + 49/Y
	7	Treatment after detn
	78	.Dump sample
51	79	.Cleanup

INDEXER ____________

DATE ____________

Form PO-1100 (1-70)

USCOMM/DC

Table 6.4 EXTRACT FROM BRITISH STEEL CORPORATION THESAURUS

```
01 COBALT
01 COHESION
01 COILERS
      02 COILERS MANDRELS
      02 COILERS WRAPPER ROLL
01 COKING
      02 COKE OVENS
            03 Coke Ovens Equipment
                  04 Coke Ovens Chambers
                  04 Coke Ovens Chargers
                  04 Coke Ovens Coke Cars
                  04 Coke Ovens Doors
                  04 Coke Ovens Flues
                  04 Coke Ovens Gas Collectors
                  04 Coke Ovens Guides
                  04 Coke Ovens Pushers
                  04 Coke Ovens Quenchers
            03 Coke Ovens Operations
                  04 Coke Ovens By Product Recovery
                  04 Coke Ovens Charging
                  04 Coke Ovens Dry Quenching
                  04 Coke Ovens Heating
                  04 Coke Ovens Pushing
                  04 Coke Ovens Quenching
            03 Coke Ovens Coal Behaviour
            03 Coke Ovens Coal Preheating
            03 Coke Ovens By Products
                  04 Ammonia Anhydrous
                  04 Ammonia Liquor
                  04 Ammonium Sulphate
                  04 Benzol
                  04 Coal Tar
                  04 Coke Ovens Gas
                  04 Napthalene
      02 COKING CONTINUOUS
      02 COKING FORMED COKE
            03 Coking Formed Coke BBF
            03 Coking Formed Coke FMC
            03 Coking Formed Coke Iniex
            03 Coking Formed Coke Lurgi
            03 Coking Formed Coke Ruhrkohl
```

searcher on a video screen, from which he can select those most relevant to his interests.

A further type of thesaurus is the diagrammatic, of which the most well-known is that developed by EURATOM (*Figure 6.1*). The linking lines between boxes often represent generic relations—for example, we could set up the hierarchy shown on p. 133.

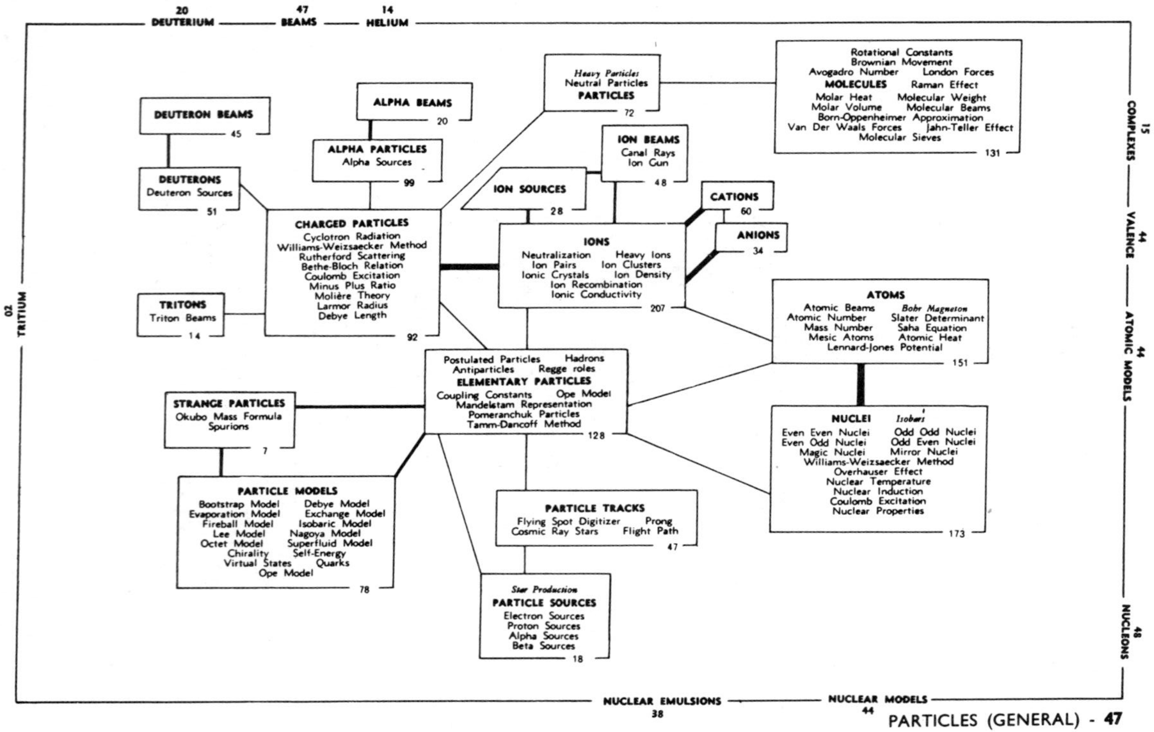

Figure 6.1 Page of EURATOM thesaurus, 2nd edition

Particles
  Molecules
  Charged particles
    Ions
      Cations
      Anions

But other lines represent relations that would be signalled as RT in an alphabetical thesaurus—e.g. between Ions and Ion sources, or between Elementary particles and Particle models: the terms in each pair come from different facets. Within each box, besides the capitalised keyword, is a set of terms more closely related to it—sometimes species of the keyword (for example, the different types of particle model), but also dependent facets (for example, the various properties of Molecules).

The diagrammatic EURATOM thesaurus is backed up by an alphabetical list, an excerpt from which is shown in *Table 6.5*.

Table 6.5 EXTRACT FROM EURATOM ALPHABETICAL THESAURUS

—Collectors	see Electrodes or Filters or Sampling or Vessels
—Colliding beams	use Beams+Collisions
Collimators	use Beam optics
Collisions	
Colloids	
Colonies	use Populations
Colony formation	use Colonies
Color	
Coloration	use Color
Colorimetry	use Color+Determination
—Columbium	use Niobium
—Columns	see Chromatography or Configuration or Mechanical structures or Thermal column
Compact reactor	use Reactors
—Compilation	see Bibliography or Programming or Tables
Compilers	use Programming

In this list terms preceded by a dash are 'forbidden', not to be used in indexing, sometimes because they are rejected in favour of a preferred synonym (Columbium, use Niobium; Colliding beams, use Beams+Collisions). Other terms are forbidden because they have several meanings (such as Columns or Compilation) and a less ambiguous term is preferred. Terms without dashes are acceptable in indexing. Those without annotation (e.g. Collisions, Colloids) figure as capitalised keywords in the diagrammatic thesaurus; the others are related terms that surround the keywords —thus Coloration and Colorimetry will both be found in the box of the keyword 'Color'. Sometimes there is a chain of such links: Colony formation—Colonies—Population.

USING HIERARCHICAL RELATIONS

The simplest use of classification in post-coordinate search has already been mentioned—the display of a section of the thesaurus (alphabetical, hierarchical or diagrammatic), either in printed form or on the video screen, to aid choice of search terms.

In the EURATOM system the hierarchical relations in the computer-held thesaurus are used to add indexing terms by the technique known as 'generic posting'. If a document is indexed by the term 'Colony formation', the term 'Colonies and Populations' will automatically be added to its indexing record. To take another example, indexing by 'Sprouting' leads to the addition of 'Plant growth', 'Plants', 'Agriculture'. Rolling expresses the situation as in *Figure 6.2*. An indexer decides that the subject of a document may be represented by terms A to G inclusive. On consulting the thesaurus, he finds that B, D, E and F are forbidden terms, thus:

—B use M
—D use P
—E use Q+R
—F use T

He therefore indexes by A, M, C, P, Q, R, T, G. The computer further consults the thesaurus it holds, and acts as follows:

M add N
C add O
P add W
Q add X
R add S
S add X
T add U
U add V
G add W

As a result, the document record now carried 15 index terms in the retrieval file. Rolling points out that the generic and 'cumulative' terms added (S, X, e.g. 'Plants', 'Agriculture') may be very

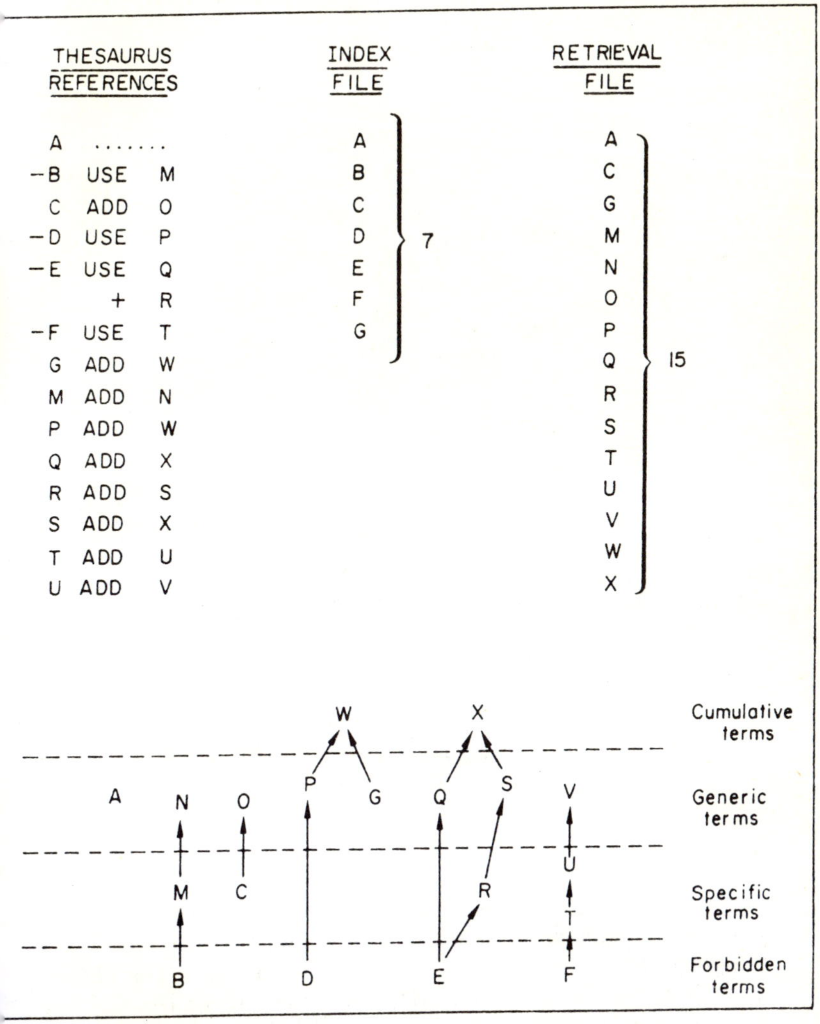

Figure 6.2 Generic posting

unlikely to figure as the topic of a specific article, but may nevertheless be very useful either for wide generic search or to delimit the broad subject field in which search is to be conducted.

In another computerised system the notation of a faceted classification is used in automatic search expansion. Eight facets are used for the field of Rock Mechanics, outlined in *Table 6.6*. Class CA, Surface structure, is shown in more detail herewith:

CA	surface structure
CA1	dams
	arch-dam
	buttress-dam
	earth-dam
	gravity-dam
	rockfill-dam
	rock-dam
CA2	slopes
	slope-angle
	slope-stability
	embankments
CA3	roads
	bridges
	footings
	highways
CA4	open-cast
	quarries
CA5	plants
	factories
	power-plants

But CD–structure material, did not need further sub-divisions in a rock mechanics environment, and so it looks as follows:

CD	structure material
	aggregates
	aluminium
	building stone
	cement
	ceramics
	clay
	concrete
	etc., according to alphabetical order

The role played by facet classification in search of this system can best be explained by an example. Suppose that a user query calls for information on 'the density of pegmatites'. The obvious initial search is for 'density AND pegmatites'. If the recall is low. the search is broadened. This is done using facet codes, for which we have Density DE3, Pegmatite AB37.

We broaden, at first, one facet at a time, thus:

DE3	density
AND AB3	density

Table 6.6 FACET STRUCTURE OF ROCK MECHANICS

A *Rocks*	*B* *Rock phenomena*	*C* *Structures*	*D* *Properties*	*E* *Measurements*	*F* *Operations*	*G* *Mental processes*
A. Age	A. Dynamic behaviour	A. Surface structures	A. Descriptive properties	A. Techniques	A. Techniques	A. Theories
B. Mineralogy, petrology	B. Gradual processes	B. Underground structures	B. Mechanical properties	B. Tools	B. Tools	B. Model studies
C. Paleontology	C. Effect of these phenomena	C. Structure components	C. Surface properties			C. Mathematical analysis
D. Structure of rocks and stratigraphy	D. Environmental factors	D. Structure material	D. Fluid flow and fluid effect properties			D. Statistical analysis
E. Tectonics and geophysics			E. Physical properties other than A. B. C. & D.			E. Economic evaluation
F. Geochemistry and chemical processes						F. Graphical presentation mapping photography
						G. Classification

	OR	
	DE	physical properties
AND	AB37	pegmatites

Next we may broaden both facets together:

	DE	physical properties
AND	AB3	igneous rocks

This strategy can be applied in the system in two ways:

(1) The interpreter can himself call for a search on one or more of the request formulations set out above.
(2) Alternatively, he may call into action an automatic search expansion routine that is built into the computer search program. The routine would run as follows:
    (a) Search for DE3 AND AB37.
    (b) Are more than three references retrieved? If not, broaden one facet, search for DE AND AB37.
    (c) Are more than three references retrieved? If not, broaden other facet, search for DE3 AND AB3.
    (d) Are more than three references retrieved? If not, broaden both facets, search for DE AND AB3.
    (e) Deliver results.

A classified thesaurus is held within the computer system of the Science Information Exchange, described by Hersey and colleagues. An excerpt is shown in *Figure 6.3.* The indexer assigns one or more specific terms (from the column labelled ULT) to each item, e.g. L–Resp–Asthma–Intract). The computer scans the index entries, and translates each into its code, in this case 581 30 100 50. The item will be selected by a search at any level of the hierarchy, such as 581 30 100 Asthma, or 581 Respiratory system.

A most complex use of classification has been developed in the computerised INSPEC services described by Field. The products of the system are (a) three separate abstracts journals, each of which is arranged according to a separate classification, (b) a separate alphabetical index to each of these journals, and (c) a retrieval file in which each bibliographic reference is linked to natural-language indexing terms. The core of the system is a 'unified classification' covering the whole range of subjects in the three journals, with a hierarchical five-digit notation. An excerpt is shown in *Table 6.7.* This is linked to the sectional classifications of the three journals—thus EEKAA Information theory occurs as heading 2.450 in journal A, as 3.410 in B, and as 6.470 in C. Many headings in the detailed unified scheme do not figure as

CODE				CAPTION	ULT
581				Respiratory System	
	05			Function, Biochemistry	
		200		Clearance, Ciliary Action	L-Resp-Clea.
		300		Control and Regulation	L-Resp-Control
		400		Diffusion of Gases	L-Resp-Diffus
		600		Mechanics, Compliance	L-Resp-Mech
		800		Surface Tension	L-Resp-Surf
		995		Other	L-Resp-Funct-O
		997		CMC	L-Resp-Funct-CMC
	30			Diseases and Conditions	
		050		Abscess	L-Resp-Abscess
		075		Atelectasis	L-Atelect
		100		Asthma	
			50	Intractable	L-Resp-Asthma-Intract
			95	Other	L-Resp-Asthma-O
		200		Bronchitis, Acute	L-Bronchit
		300		Chronic Respiratory	
			20	Bronchiectasis	L-Bronchiect
			40	Emphysema	L-Emphysema
			97	CMC	L-Chronic-Resp-CMC
		400		Hyaline Membrane	L-Hyaline
		500		Irritants, Inhaled	
			50	Gases	L-Inhal-Gas
			80	Particulates	L-Inhal-Part
			95	Other	L-Inhal-O
		700		Pneumonia, Pneumonitis	L-Pneumonia
		725		Pulmonary Alveolar Proteinosis	L-Proteinosis
		750		Pulmonary Edema	L-Pul-Edema
		775		Pulmonary Fibrosis	L-Pul-Fib
		995		Other	L-Resp-Dis-O
		997		CMC	L-Resp-Dis-CMC

Figure 6.3 Science Information Exchange: extract from classified thesaurus

headings in the journals. Next, each class in the unified scheme is linked to more specific indexing terms—for example, class GMGAA Mesons has associated with it Meson magnetic moment, Meson mass, Meson resonances, Meson spin, Pions. If an item is indexed by the term 'Meson resonances', it is thereby assigned by computer to class GMGAA, which in turn links it to, say, heading 10.45 in journal A.

Further, thesaurus relations between specific terms have been built up, as shown in *Table 6.8*. Thus, the term 'Meson resonances' is used for (UF) Vector mesons; it has narrower terms (NT) such as 'Omega mesons', broader terms (BT) such as 'Bosons', a top term (TT) 'Elementary particles', and related terms (RT) such as 'Mesons'; and it is classed as we have seen GMGAA (the sixth letter of the class symbol is a computer check digit). From the thesaurus relations can be derived headings for the printed indexes, thus:

'A UF B' becomes 'B see A'
'A NT C' becomes 'A see also C'

'A RT D' becomes 'A see also C'
and 'C see also A'
'A BT E' becomes 'E see also A'

From the BT/NT relations can also be prepared an alphabetico-classed listing of specific terms, as shown in *Table 6.9*, and either this or the relational thesaurus is available on display to aid search of the retrieval file. To serve the various functions of the INSPEC system there is therefore a complex use of classification.

SYNTACTIC RELATIONS

As already noted, in a straightforward post-coordinate system syntactic relations, between the terms representing a topic, are omitted. Since the terms are not pre-coordinated, there is no

Table 6.7 UNIFIED INSPEC CLASSIFICATION

EEAAAS **OPTICS** *(for instrumentation see BGHAAC; for properties of gases and of liquids and solids see LCTAAH and RVAAAQ respectively)* *a 2.400*

EECAAH	**Propagation and transmission in homogeneous and isotropic media** *a 2.410*
EECACM	*General*
EECAER	*Wave front and ray tracing*
EECAGW	*Edge and boundary effects*
EECAKC	*Scattering and reflection from extended bodies*
EECAMG	*Absorption*
EECARQ	*Coherence*
EECAZF	*Other and miscellaneous*
EEEAAY	**Propagation and transmission in homogeneous and anisotropic media** *a 2.420*
EEGAAN	**Propagation and transmission in inhomogeneous media** *a 2.440*
EEGACS	*General*
EEGAEX	*Wave front and ray tracing and beam spread in random, turbulent media*
EEGAGB	*Coherence in random, turbulent media*
EEGAKH	*Scattering from haze, fog, dust, etc*
EEGAZL	*Other and miscellaneous*
EEKAAL	**Information theory** *a 2.450 b 3.410 c 6.470*
EEKACQ	*General*
EEKAEV	*Fourier transform optics*
EEKAGZ	*Modulation transfer functions*
EEKAKF	*Optical communications*
EEKAMK	*Pattern recognition*
EEKAZJ	*Other and miscellaneous*
EEMAAB	**Holography** *(for techniques and instruments see BGHRAJ)* *a 2.460*
EEMACF	*General and theoretical*
EEMAEK	*Image characteristics*
EEMAGP	*Photographic problems*
EEMAKW	*Applications*
EEMAZZ	*Other and miscellaneous*
EERAAG	**Image formation** *a 2.480*
EERACL	*General*
EERAEQ	*Theory*
EERAGV	*Aberrations*
EERAKB	*Resolution*
EERAZE	*Other and miscellaneous*

Table 6.8 EXTRACT FROM INSPEC THESAURUS

meson production
- NT kaon production
  - meson photoproduction
  - pion production
- BT hadron production
- TT hadron production
- RT mesons
- CC GFAAAH GKAAAF

meson reactions
- USE meson interactions

meson resonances
- UF kaon resonances
  - resonances, meson
  - vector mesons
- NT omega mesons
  - phi mesons
  - rho mesons
- BT bosons
  - hadrons
- TT elementary particles
- RT mesons
- CC GMGAAA

meson scattering
- NT kaon scattering
  - meson-baryon scattering
  - meson-meson scattering
  - pion scattering
- BT hadron scattering
- TT elementary particle scattering
- RT meson interactions
  - mesons
  - nuclear reactions and scattering due to mesons
- CC GKCAAW GKEAAL

meson spin and parity
- BT parity
  - spin
- TT parity
  - spin
- RT mesons
- CC GMGAAA

unique sequence among them. A regulated sequence is of value in a visually searched pre-coordinate index, not only to ensure consistency, but also to produce an intelligible string of terms. Examples of methods of achieving this have been discussed in Chapters 2 and 3. Syntactic relations have also been introduced into post-coordinate systems.

In a coordinated string of terms, if each term falls within a defined category, there is an implied relation between successive terms.

Table 6.9 EXTRACT FROM INSPEC ALPHABETICO-CLASSED VOCABULARY

elementary particle decay
. electromagnetic decays
. hadron decay
. . baryon decay
. . . baryon hadronic decay
. . . baryon leptonic decay
. . . hyperon decay
. . meson decay
. . . kaon decay
. . . . kaon hadronic decay
. . . . kaon leptonic decay
. . . meson hadronic decay
. . . . kaon hadronic decay
. . . meson leptonic decay
. . . . kaon leptonic decay
. . . pion decay
. leptonic decays
. . baryon leptonic decay
. . meson leptonic decay
. . . kaon leptonic decay
. . muon decay
. semi-leptonic decays

elementary particle interactions
. alpha-particle interactions
. . alpha-particle absorption
. cosmic ray effects and interactions
. . nuclear reactions and scattering due to cosmic rays
. deuteron interactions
. elementary particle electromagnetic interactions
. elementary particle strong interactions
. . duality and dual models
. . . Veneziano model
. . elementary particle inclusive reactions
. . hadron classification schemes
. elementary particle weak interactions
. hadron interactions
. . baryon interactions
. . . baryon-baryon interactions
. . . . hyperon-nucleon interactions
. . . . nucleon-nucleon interactions
. . . . . proton-proton interactions
. . . hyperon interactions
. . . . hyperon capture
. . . . hyperon-deuteron interactions
. . . . hyperon-nucleon interactions

For example, the relation between Thing and Constituent, or between Thing and Property, is what Farradane has called 'appurtenance', and that between Thing and Operation or Process he has called 'reaction'. In general, if T1 is a term within category or facet F1, a string of such terms may be represented as F1T1—F2T2—F3T3. If R12 is the relation between F1 and F2, we could transform this string into T1—R12—T2—R23—T3, which represents Farradane's solution. But in a post-coordinate system T1, T2 and T3 are separated. Relations between them can be signalled (a) by attaching a 'role indicator' to each, which may also be represented as F1T1, F2T2, F3T3, or (b) by attaching directed relations, thus T1R12, R21T2R23, R32T3, showing that T2 is linked to both T1 and T3.

The first solution was adopted in a number of post-coordinate systems, the most widely used set of roles being those developed by the US Engineers Joint Council (*Table 6.10*).

Some examples of the use of these roles are as follows:

'In which articles has heat transfer been the *primary topic* of consideration?' HEAT TRANSFER–8.
'What information on butane as a *raw material* in the collection?' BUTANE–1.
'Do we have any articles which discuss *processes* or *operations* where isopentane is produced?' ISOPENTANE–2.
'Which articles mention the use of carbon tetrachloride as a *solvent?*' CARBON KETONE–3.
'Do we have any articles where pumps have been mentioned for possible *application* of our products?' PUMPS–4.
'Which articles mention the use of carbon tetrachloride as a *solvent?*' CARBON TETRACHLORIDE–5.
'In which articles is temperature discussed as an *independent variable?*' TEMPERATURE–6.
'Is there any information on conversion *being affected* by other operating variables?' CONVERSION(CHEMICAL)–7.

Examples of searches requiring two or more term-roles:

'Is there any information on the synthesis of benzene from cyclohexane?' SYNTHESIS–8, CYCLOHEXANE–1, BENZENE–2.
'Have storage tanks ever been cleaned by sandblasting?' CLEANING–8, TANKS–9, STORAGE–4, SANDBLASTING–10 (role 9 is used here to show passive receiver of an action).
'How does the concentration of tetramethyl lead affect the road octane number of gasoline?' CONCENTRATION–6, TETRAMETHYL

Table 6.10 ROLES DEVELOPED BY ENGINEERS' JOINT COUNCIL

8
The primary topic of consideration is; the principal subject of discussion is; the subject reported is; the major topic under discussion is; there is a description of

1
Input; raw material; material of construction; reactant; base metal (for alloys); components to be combined; constituents to be combined; ingredients to be combined; material to be shaped; material to be formed; ore to be refined; sub-assemblies to be assembled; energy input (only in an energy conversion); data and types of data (only when inputs to mathematical processings); a material being corroded

2
Output; product, by-product; outcome, resultant; intermediate product; alloy produced; resulting material; resulting mixture or formulation; material manufactured; mixture manufactured; device shaped or formed; metal or substance refined; device, equipment, or apparatus made, assembled, built, fabricated, constructed, created; energy output (only in an energy conversion); data and types of data (only as mathematical processing outputs)

3
Undesirable component; waste; scrap; rejects (manufactured devices); contaminant; impurity, pollutant, adulterant, or poison in inputs, environments, and materials passively receiving actions; undesirable material present; unnecessary material present; undesirable product, by-product, co-product

4
Indicated, possible, intended present or later uses or applications. The use or application to which the term has been, is now, or will later be put. To be used as, in, on, for, or with; for use as, in, on, for, or with; used as, in, on, for, or with; for later use as, in, on, for, or with

5
Environment; medium; atmosphere; solvent; carrier (material); support (in a process or operation); vehicle (material); host; absorbent adsorbent

6
Cause; independent or controlled variable; influencing factor; 'X' as a factor affecting or influencing 'Y', the 'X' in 'Y' is a function of 'X'

7
Effect; dependent variable; influenced factor; 'Y' as a factor affected or influenced by 'X'; the 'Y' in 'Y' is a function of 'X'

9
Passively receiving an operation or process with no change in identity, composition, configuration, molecular structure, physical state, or physical form possession such as when preceded by the preposition *of, in,* or *on* meaning possession; location such as when preceded by the prepositions *in, on, at, to* or *from* meaning location; used with months and years when they locate information (not bibliographic data) on a time continuum

10
Means to accomplish the primary topic of consideration or other objective

0
Bibliographic data, personal names of authors, corporate authors and sources, types of documents, dates of publication, names of journals and other publications, other source-identifying data, and adjectives.

Some examples of the use of these roles are as follows:

LEAD–9, ROAD OCTANE NUMBER–7, GASOLINE–9 (role 9 is used here in the possessive sense; role 6–7 implies EFFECT–8).

The main example of the use of directed syntactic relations is the system developed by Perry and Kent. To express a topic, their WRU (Case Western Reserve University) system constructed a syntactic string of terms. The terms themselves were coded, but this aspect will not be considered here. Using natural-language terms, a simple phrase string might be KOV. *cermet*, KWV.KAP. *brittleness*, KAL, *microstructure*, where the three letter codes represent directed relations. There are four role indicators in the phrase:

KOV = *property given for*
KWV = *property given*
KAP = *property influenced*
KAL = *influenced by*

These indicators express two binary relations among the three terms: (i) brittleness is a property of cermets, (ii) brittleness is influenced by microstructure. Each binary relation must be expressed by adding an indicator to *each* term in the relation, thus:

(i) KOV *cermet*, KWV. *brittleness*
(ii) KAP *brittleness*, KAL. *microstructure*

The encoding is shortened by combining the two to give

(iii) KOV. *cermet*, KWV.KAP. *brittleness*,
KAL. *microstructure*

It is necessary to assign a role indicator to each of the related terms, since juxtaposition is not meaningful in the code.

All phrases in an encoded abstract are built up in this way, by combining role indicators and terms. About 20 role indicators are provided. Nine of these can be applied to terms denoting *material:* KOV (*property given for*), KEJ (*material processed*), KAJ (*starting material*), KUJ (*component*), KWJ (*product*), KQJ (*by means of*), KAD (*machine* or *device*), KAG (*subassembly*) and KAL (*influenced by*). Four of them can be applied to terms denoting *property:* KWV (*given*), KUP (*determined*), KAP (*influenced*) and KAL (*influencing*). Three can be applied to *processes:* KAM (*process*), KXM (*negation*) and again KAL. One represents an attendant *condition* (KAH), and there are eight miscellaneous indicators: KIS (*location*), KWB (*direction from*), KWC (*direction to*), KAB (*field*), KIT (*time occurring*), KIG (*geographical location*), KIB (*organisation involved*) and KEP (*person involved*).

By combining terms and role indicators into sub-phrases, these into phrases, these again into *sentences* and *paragraphs*, the encoded abstracts are built up. An example is as follows:

Theme: The fabrication by welding of the body of the Dyna-Panhard 54 automobile, using aluminium–3.5% magnesium alloy.
Encoded: KAJ. alloy, KUJ. Al, Mg * per cent *, NLMT 3/5.,—KAM. fabrication, KAM. welding,—KWJ. body * Dyna-Panhard automobile * 54, KUJ. alloy, KUJ. Al, KUJ. Mg * per cent *, NLMT. 3/5.

There are three basic binary relations involved: (1) a car body is fabricated, (2) the body consists of an alloy, (3) the fabrication is by welding. The third relation is really ternary: (3) the fabrication is by welding the alloy. In order to express this ternary relation, the encoded abstract repeats the coding for the alloy to give the result—alloy, fabrication/welding, body, alloy.

Such developments are theoretically interesting, and the use of relational terms in both pre- and post-coordinate systems continues to receive attention—as recently reviewed by Coates—but there is no strong evidence that they are necessary for effective retrieval in post-coordinate systems.

BIBLIOGRAPHY

E. J. COATES. 'Some properties of relationships in the structure of indexing languages'. *J. Docum.*, 1973, **29,** 390–404

EURATOM Thesaurus. Brussels, 2nd ed., 1967

B. J. FIELD. *Development of an Integrated Indexing and Classification System*. London, 1972

A. GRALEWSKA. 'The design of an information system for the literature of Rock Mechanics'. *J. Docum.*, 1968, **24,** 197–209

D. F. HERSEY and others. 'Conceptual indexing and retrieval of current research records'. *Method. Inform. Med.*, 1968, **7,** 172–87

S. J. MARTINEZ and D. P. HELANDER. 'The development and maintenance of a specialised, controlled vocabulary thesaurus'. *Proc. Amer. Docum. Inst.*, 1968, **5,** 279–83

J. G. MULVIHILL and E. H. BRENNER. 'Faceted organisation of a thesaurus vocabulary'. *Proc. Amer. Docum. Inst.*, 1966, **3,** 175–83

M. N. PATTEN. 'Storage and retrieval of information using VDU's on-line'. Paper at *First European Congress on Documentation Systems and Networks, Luxembourg 1973*

J. W. PERRY and A. KENT (eds.). *Tools for Machine Literature Searching*. New York, 1958

L. N. ROLLING. 'Compilation of thesauri for use in computer systems'. *Inf. Stor. Retr.*, 1970, **6,** 341–50

G. SALTON. *Automation Information Organisation and Retrieval*. New York, 1968

APPENDIX A

HISTORICAL ASPECTS OF THE CLASSIFICATION OF SCIENCE

'The great inherent defect of present classification systems', wrote Lund and Taube some years ago, 'is their explicitly stated or implied claim to permanence. . . . Knowledge is supposed to have reached the stage where a comprehensive and fairly definite classification can be made to include it all under a definite number of main headings.' In reality, history presents a series of cultural epochs. Each is a span of years within which knowledge presents a more or less unified structure which can be expressed in a classification, but each new epoch requires a new classification. 'All that any system can hope to do is to classify accurately and completely knowledge at any given stage of its development.'

The most casual survey of the history of thought confirms the view. Each age has its own 'flavour', it studies particular problems, its methods of study are characteristic, it exhibits a special mode of thought which serves as a framework to its theories, the dominant intellectual interests are unique. An organisation of knowledge which fits the philosophical temper of the Greeks does not suit the mythological cosmologies of Babylon, the rushing life of the Renaissance, or the industrialised nineteenth century. Indeed, the better fitted a classification is to a given epoch, the less suitable will it be for any other epoch.

Knowledge ceaselessly advances. With scientific knowledge, above all others, that advance is ceaselessly accelerating. We might expect therefore, the cultural 'epochs' will have succeeded each other with increasing frequency, particularly since the Renaissance. Lund and Taube suggest this, and give a tentative list of periods in the history of civilisation (dates added):

to 600 B.C.	Early civilisation
600–200 B.C.	Hellenic
300 B.C.–A.D. 400	Hellenistic and Roman
A.D. 400–1300	Mediaeval and Arab
A.D. 1300–1600	Age of transition

A.D. 1600–1700	Seventeenth century
A.D. 1700–1850	Age of physical science
A.D. 1850–1914	Age of 'progress'
A.D. 1914–	From World War I

It is of some interest to consider the history of the classification of science with this in mind. The use of the term 'science' to denote the study of the material world, inorganic and organic, is relatively recent. The classical term for the study of 'nature' was *physica*, the Latin translation of the title of a work by Aristotle. Though he presented no formal classification of knowledge, the meaning of *physica* can be derived from a study of his works.

Aristotle in effect defined Physics as the study for its own sake (not for practical or productive ends) of things which have a separate, concrete existence (unlike the world of numbers) and are subject to change (unlike the deathless gods). The term thus embraced the whole of what we now call natural science except the 'mathematical' sciences, in which 'number' reigned supreme—Arithmetic, Geometry, Astronomy and Harmonics.

Aristotle's system of the sciences—which will be described in more detail later—was the framework of knowledge for nearly two thousand years, from 300 B.C. to A.D. 1600 and even beyond. It began to break down, however, during the late Middle Ages and the Renaissance, and in the eighteenth century was finally abandoned. The subsequent history of classification has been the attempt to find a new rational system of the sciences to replace that of Aristotle. In this chapter I shall trace the details of this historical development and try to assess its significance for us today. Outlines of the main schemes discussed are given at the end of this chapter.

THE RISE AND FALL OF ARISTOTLE'S SYSTEM

For 2000 years the classification of knowledge mainly followed the pattern laid down by Aristotle. Although his synthesis was only imperfectly known during the Dark Ages in Europe, and began to show signs of strain towards the close of the Middle Ages, it remained the basic framework of knowledge until almost the end of the seventeenth century. The origin, dominance and subsequent decline of the Aristotelean system will now be briefly sketched.

The first writers to attempt a survey of contemporary knowledge were the philosophers of Greece. In the *Republic* Plato described the sciences as instruments of that culture which was to make the philosopher a fit ruler of the state. The highest study he called 'dialectic', and leading up to it were four sciences: Arithmetic, Geometry, Astronomy and Harmonics. To Plato, Arithmetic was not the art of computation, but the abstract study of number. Astronomy and Harmonics were likewise the abstract mathematical study of celestial motions and musical sounds. Although in the *Timaeos* Plato discussed many other physical and biological subjects, it was only the four studies mentioned which he considered to be 'sciences'.

Aristotle took a wider view. He divided knowledge into three parts—the Theoretical, which aims at knowledge for its own sake; the Practical, which seeks knowledge as a guide to conduct; and the Productive, which aims to be used in making things useful or beautiful. Practical knowledge included Ethics, Politics, Economics and Rhetoric, while Productive knowledge covered Poetics and the Arts. It is therefore mainly Theoretical knowledge which is akin to the modern conception of science. Aristotle regarded Logic not as a substantive science, but as a study preparatory to all the sciences.

Theoretical knowledge Aristotle further divided into Theology or Metaphysics, which deals with things that have a separate existence and are unchangeable; Mathematics, which studies things that are unchangeable but have no separate existence; and Physics, which deals with changeable things that have a separate existence. The content of Aristotle's Mathematics included Plato's four sciences—Arithmetic, Geometry, Astronomy and Harmonics—but added to them Optics and Mechanics. His Physics, however, incorporated fields of knowledge which Plato largely neglected. It covered (1) the 'first causes' of nature, i.e. the nature of change and movement in general, (2) the order and motion of the stars (Physical Astronomy) and of terrestrial bodies, (3) 'coming into being' and 'passing away' in general, (4) Meteorology, which included Mineralogy and the study of chemical change, and (5) Biology, Botany and Zoology.

MEDIAEVAL SCIENCE

Rome added little to science, and after its decline the heritage of Greek science passed mainly to the Arabic-speaking peoples. The Latin culture of Europe at first followed the Platonic tradition of science. The 'seven liberal arts' became the disciplines to be studied preparatory to the higher studies of theology, metaphysics and ethics. Of these seven arts, the *trivium* covered Logic, Grammar and Rhetoric. The *quadrivium* coincided with Plato's four sciences.

The new content of Physics which Aristotle had brought into science was lost to view.

An early exposition of the quadrivium was that of Martianus Capella, in his *Satyricon* written about A.D. 470. In the seventh century Isidore of Seville added Medicine to the list, and in the ninth Rabanus Maurus added Mechanical Arts. No further extension of the scope of science occurred in Europe until the recovery of Greek science from the Arabs had begun.

During the period 750–900 the scholars of Jundishapur and Baghdad had made available to the Arabic-speaking world the major part of Greek science: the whole of its mathematics, mechanics, astronomy, mathematical geography, medicine and zoology, though not all of its botany. Early in the tenth century the Turkestan philosopher al-Farabi (Alpharabeus) prepared a classification of the sciences. Like Aristotle, he first treated Logic separately, then outlined seven parts of Mathematics: Arithmetic, Geometry, Music, Astronomy, Optics, Statics and Mechanical Devices. His enumeration of the parts of Physics was also based directly on that of Aristotle. Three centuries later, the Persian Nasir al-din al-Tusi introduced new sciences into the Aristotelean classification. In Mathematics he included Algebra, which had been developed by Hindu and Arab scientists, and to Physics he added Medicine, Astrology and Agriculture.

The translation of scientific texts from Arabic into Latin began in the eleventh century in Italy and Spain, and was largely completed during the following century. The absorption of Graeco-Arab science by Western Europe was followed by the construction of classifications. The most important mediaeval writer was Hugh of St. Victor. He divided knowledge into four parts: Logic (including Grammar and Rhetoric, i.e. the old trivium); the practical sciences of Ethics, Politics and Economics; theoretical science; and mechanics. Theoretical science Hugh divided, like Aristotle, into Theology, Mathematics (the quadrivium) and Physics (in Aristotle's sense). The novelty of Hugh's classification lay in the introduction of seven 'mechanical arts': Cloth Manufacture, Armour-making, Navigation, Agriculture, Hunting, Medicine and Theatrical Science. In the thirteenth century Robert Kilwardby of Canterbury suggested that the list be amended by the replacement of Navigation, Hunting and Theatrical Science by Food Science, Architecture and Commerce.

Roger Bacon planned a vast encyclopaedia arranged in four parts: (1) Grammar and Logic; (2) Mathematics—the quadrivium; (3) Natural Science; and (4) Metaphysics and Morals. The third part was to be sub-divided into the Aristotelean Physics and a number of 'special sciences'—Optics, Astrology, Statics, Alchemy, Agriculture and Medicine.

Viewing the mediaeval classifications of knowledge as a whole, we may consider what developments they made as compared with the Aristotelean synthesis. 'Science', as we understand it today, was still in no way separated from other departments of knowledge. The abstract mathematical sciences recognised by Plato still hung together as a unity. Aristotelean Physics was preserved almost unchanged. But the growth of knowledge had opened up a number of new fields, and various methods were tried of incorporating them into Aristotle's scheme. Optics and Mechanics, for example, were placed by the Arab writers in Mathematics, as Aristotle had suggested, but without reservations. Aristotle himself had called these studies (as well as Astronomy and Harmonics) 'the more physical parts of mathematics'. Nasir al-din followed him in considering Optics, Mechanics (and Algebra) to be 'derivative' parts of Mathematics rather than its 'elements'. Roger Bacon took Optics and Statics out of Mathematics altogether and included them with the 'special sciences'.

The applied sciences—Medicine, Agriculture and so on—which had been ignored by Aristotle, were tacked on to Physics by Nasir al-din, but Roger Bacon and Hugh of St. Victor put them in a group by themselves, replacing Aristotle's 'productive sciences'.

Although, therefore, the basic division of science into Mathematics and Physics still remained, by the thirteenth century in both Arab and European cultures it was becoming increasingly difficult to incorporate the new scientific and industrial knowledge into this framework. A new class of 'derivative' or 'special' or 'mechanical' sciences was being created.

THE RENAISSANCE

The next stage in the development of science, which I here loosely call the Renaissance, covers the period 1400–1700. The breakdown of the Aristotelean synthesis, begun in the later Middle Ages, was completed during this period.

The classification of Conrad Gesner's *Pandectarum* (1548) was basically that of Aristotle. His 'substantial' sciences included Natural philosophy (which in its sub-divisions exactly followed Aristotle's Physics), Metaphysics, Ethics, Economics, Politics, Law, Medicine and Theology. The necessary 'preparatory' sciences included the trivium and quadrivium. 'Ornamental' preparatory sciences included a diverse list of mechanical and other arts. To those of Robert Kilwardby and Hugh of St. Victor, Gesner added Metallurgy, Printing, Shipbuilding, Chemistry, Pharmacy and others.

Francis Bacon in the *Advancement of Learning* retained Aristotle's basic dichotomy, but gave it a new interpretation. Aristotle's Mathematics became part of Natural Philosophy. This included the study of causes (the first part of Aristotle's Physics). Mathematics itself was sub-divided into 'pure' (Arithmetic and Geometry) and 'mixed' (Optics, Music, Astronomy, Architecture and Engineering)—a division reminiscent of Nasir al-din's. Bacon classed the bulk of Aristotle's Physics as Natural History, together with the Mechanical Arts and Agriculture. Medicine he included in Human Philosophy.

In his writings, Bacon frequently attacked Aristotle. He denied that knowledge could be divided into theoretical, practical and productive, and instead suggested a division into history, poetry and philosophy. The first of these was a 'descriptive' science based on memory, furnishing the data on which 'speculative' philosophy, based on reason, went to work. But the 'production of effects'—the practical and productive aspects of knowledge—followed directly on the knowledge of causes which philosophy provided, and therefore Bacon included the 'production of effects' as an 'operative' part of Natural Philosophy.

The point to be noted, however, is that despite these theoretical disagreements with Aristotle, the basic groupings of Aristotelean Physics and Mathematics appeared almost unchanged in Bacon's classification. It was only the reasons given for the grouping which were changed. For Aristotle, Physics dealt with changeable, independently existing things, and Mathematics with unchangeable things which had no separate existence; for Bacon, Natural History dealt with the variety of things, and Mathematics with the fixed, formal causes of those same things. The same basic grouping appeared in the classification put forward by Thomas Hobbes in *Leviathan* (1651), but once again the interpretation was changed. To Hobbes, Physics dealt with the qualitative aspects of things, and included Meteorology, Physical Astronomy, Mineralogy, Botany, Zoology, Optics and Music; the quantitative studies included Geometry, Arithmetic, Mathematical Astronomy and Geography, Mechanics, Engineering, Architecture and Navigation. Though this differed in one detail from the Aristotelean scheme (Optics and Music were transferred to Physics), it preserved it in principle. Even as late as 1718, the Aristotelean division was still used by Leibniz, who divided Natural Philosophy into physical (Physics, Chemistry, Mineralogy, Botany, Zoology) and mathematical (Mathematics, Astronomy, Mechanics).

By the beginning of the eighteenth century, however, the Aristotelean scheme was generally dropped. Its final collapse was followed by a period in which for practical arrangements reasoned classificatory groupings were no longer made, though philosophers such as Hegel continued to devise schemes. Instead, a simple enumeration of basic disciplines occurred. This method was followed in the encyclopaedia of Johann Alsted (1630), who listed ten 'theoretical' sciences (Metaphysics, Pneumatics, Physics, Arithmetic, Geometry, Cosmography, Astronomy, Geography, Optics, Music) followed by a number of 'arts' (Mechanical, Agriculture, Metallurgy, etc.). The library classification of Prosper Marchand (1704) followed a similar pattern.

By the eighteenth century, therefore, the old Aristotelean classification of knowledge had broken down, but it was not clear what was to take its place. During the nineteenth century, attempts were made to construct a new rational basis for grouping the sciences. The attempts were as varied as they were numerous, and the period can only be described as one of intellectual confusion, in which a succession of contradictory systems competed for attention.

TRADITIONAL DISCIPLINES

In 1725 John Lowthorp published the first volume of an abridgement of the *Philosophical Transactions* of the Royal Society, disposing the subject matter under a number of headings—a simple enumeration—which provide an indication of the fields of study which were recognised at that time:

Mathematics	Architecture	Magnetics
Surveying	Shipbuilding	Botany
Optics	Perspective	Agriculture
Astronomy	Music	Zoology
Mechanics	Physiology	Anatomy
Acoustics	Meteorology	Medicine
Hydromechanics	Pneumatics	Pharmacy
Geography	Hydrology	Chemistry
Navigation	Mineralogy	

The most interesting feature of the list is the final disintegration of the old Aristotelean Physics into a number of separate disciplines —from Physiology to Chemistry. At least two new disciplines unfamiliar to Aristotle had appeared: Magnetics and Pharmacy. During the nineteenth century, a number of other disciplines were added to the list of sciences—for example, Heat, Electricity, Geology,

Crystallography, Palaeontology, Atomic Physics, Embryology, Bacteriology.

These traditional disciplines henceforward formed the 'raw material' out of which classifications of science were constructed. The problem on which nineteenth-century classification concentrated was to discover how these disciplines, traditional fields of study, were to be arranged. Two approaches were made to this question: (a) Could the sciences be grouped, and (b) if grouping proved difficult, could they be arranged in a serial order?

DICHOTOMIES

The distinction between what are now called the pure and applied sciences had already emerged in mediaeval classification, as we have seen—the 'derivatives' of Nasir al-din's Physics, and the Mechanical Arts of Hugh. Alsted also separated the 'theoretical' sciences from the 'arts'. The same separation was made in the nineteenth century by Lord Lindsay, W. D. Wilson, Alexander Bain, Melvil Dewey, Thomas Masaryk, Edmond Goblot and others. Not all classifiers, however, took this step. Thomas Young and Neil Arnott kept technologies with the sciences on which they depended, a practice later followed in various ways by J. D. Brown, E. W. Hulme, S. R. Ranganathan, and H. E. Bliss.

Throughout the century, apart from the division just described, classifiers attempted to divide the sciences into two groups. Already they had before them the examples of Francis Bacon (speculative and descriptive) and Hobbes (quantitative and qualitative). For Coleridge, the sciences were either 'pure' (Grammar, Logic, Rhetoric, Mathematics, Metaphysics) or 'mixed'. Schopenhauer's similar groups were called 'pure' and 'empirical', Wilhelm Wundt called them 'formal' and 'empirical', Goblot 'mathematical' and 'theoretical', and the St. Louis Congress of Arts and Sciences (1904) 'normative' and 'physical'. Karl Pearson made a similar division of the sciences into 'abstract' and 'concrete'.

These last two terms had been brought into classification by August Comte (1830). 'We must distinguish,' he wrote, 'between the two classes of Natural Science—the abstract, or general, which have for their object the discovery of laws which regulate phenomena in all conceivable cases; and the concrete, particular, or descriptive, . . . whose function it is to apply those laws to the actual history of existing beings.' Herbert Spencer used this distinction to divide the sciences into three groups—the abstract, 'concerned with ideal or unoccupied forms of relations'; the abstract-concrete, treating of 'realities as manifested in their different modes, when these are artificially separated from one another'; and the concrete, which 'have for the terms of their relations, those heterogeneous combinations of forces that constitute actual phenomena'. H. E. Bliss has given much attention to what he describes as this 'abstract–concrete blunder'—a distinction which subsists in thought, but which cannot be successfully used to divide the sciences.

The terms were also used by other nineteenth-century classifiers, to describe groupings which did not coincide with those of Pearson. For example, Alexander Bain listed as 'fundamental or abstract' sciences Logic, Mathematics, Physics (in the modern sense), Chemistry, Biology and Psychology, and called 'dependent or concrete' Mechanics, Acoustics, Astronomy, Botany, Zoology, Meteorology and others. Masaryk's use of the terms was very similar.

Other dichotomies were also introduced in order to sub-divide the main groups further. Schopenhauer classed Mechanics and Physiology as 'general' empirical sciences, Astronomy and Zoology as 'special'. For Wundt, Dynamics was a 'general' but Physics and Chemistry 'special' natural processes. Goblot contrasted 'general' cosmology (Physics) with 'applied' (Chemistry, Astronomy, Geology, etc.). Wilson considered Geology, Geography, Mineralogy and Chemistry to be 'exact' rather than 'pure' theoretical sciences, but Pearson called them 'synoptic' rather than 'exact' inorganic sciences.

Throughout the century, therefore, a series of dichotomies was employed: speculative and descriptive, pure and mixed, pure and empirical, formal and empirical, normative and physical, abstract and concrete, fundamental and dependent, general and special, general and applied, pure and exact, exact and synoptic. These terms were used in repeated—and contradictory—attempts to group the sciences. The fact that the attempt was so often repeated, and that earlier terminology was so often discarded as unsuccessful, leads one to suspect that the aim itself was false, and justifies the characterisation of this period as one of intellectual confusion.

The plain fact is that there are *no* sciences which are wholly speculative, pure, formal, normative, abstract, fundamental, general or exact—or the opposite. As Bliss has stated, 'the sciences are all partly abstract and theoretical, and partly concrete and descriptive'. The attempts of these nineteenth-century classifiers to find a new basic for dividing up the sciences into groups—to replace the Aristotelean synthesis—led nowhere.

SERIAL ORDER

Parallel with the attempts to find a rational basis for dividing the sciences into groups there ran another stream of thought—the endeavour to establish a *serial order* of the recognised main disciplines. In part, this was a development of the abstract–concrete or general–special dichotomy put forward by Comte. He grouped the sciences in the order: Mathematics, Astronomy, Physics, Chemistry, Biology. This order, he declared, was one of decreasing generality and increasing complexity. Each science in the series was alleged to depend on those that preceded it in the order, but not on those that followed it. Spencer's order of sciences was rather different (Logic, Mathematics, Mechanics, Physics, Chemistry, Astronomy, Geology, Biology), but he developed the idea of serial dependence. Each science was *instrumental* with respect to those that followed it in the series, and supplied subject-matter to those that preceded it.

These ideas of serial order undoubtedly expressed an important truth, the fact of dependence, but did no more than state it. They did not reveal the essential *natures* of, say, Physics and Chemistry, which led to Physics being a tool in the study of Chemistry, and to Chemistry providing subject-matter for physical investigation. Attempts to specify the essential nature of each science were made by William Whewell and P. E. Dove. Whewell maintained that the classification of the sciences 'depends neither upon the faculties of the mind to which the separate parts of our knowledge owe their origin, nor upon the objects which each science contemplates; but upon . . . the *Ideas* which each science involves'. Each science was constructed about one central idea, e.g. Geometry on the idea of 'space', Arithmetic on 'number'. Moreover, each science involved, not only its own central idea, but also all the ideas of the sciences which preceded it in the order. Dove presented a similar thesis, calling the central idea the 'object-noun' of a science. Thus Logic depended on 'identity', Arithmetic was Logic applied to 'number', Algebra was Logic and Arithmetic applied to 'quantity', and so on.

In this way each author built up a serial order of sciences, of increasing complexity. Since any science included all the 'ideas' or 'object-nouns' which preceded it, it necessarily supplied subject-matter to sciences earlier in the order, which in turn might act as instruments to investigate it. The serial orders put forward by Whewell and Dove were as follows:

Whewell		*Dove*	
Idea	*Science*	*Object-noun*	*Science*
Space	Geometry	Identity	Logic

Whewell		*Dove*	
Idea	*Science*	*Object-noun*	*Science*
Number	Arithmetic	Number	Arithmetic
Sign	Algebra	Quantity	Algebra
Limit	Differentials	Space	Geometry
Motion	Formal astronomy	Force	Statics
Force	Statics	Motion	Dynamics
Matter	Dynamics	Weight	Mechanics
Exertia	Hydrostatics	Sound	Acoustics
Fluid pressure	Hydrodynamics	Light	Optics
Medium	Acoustics	Heat	Thermotics
Qualities	Optics, thermotics	Magnetic force	Magnetism
Polarity	Electricity	Electric force	Electrons
Substance	Chemistry	Affinity	Chemistry
Symmetry	Crystallography	Organisation	Physiology
Likeness	Mineralogy		
Degrees of likeness	Systematic botany and zoology		
Natural affinity	Anatomy		
Organisation	Biology		
Emotion	Psychology		
Historical causation	Geology Ecology		

A further essay in distinguishing sciences by their central concepts was made by Wilhelm Ostwald to give the following series:

Concept	*Science*
Order	Formal sciences
Manifold	Logic
Quantity	Mathematics
Space	Geometry
Motion	Kinematics
Energy	Physical sciences
	Mechanics
	Physics
	Chemistry
Life	Biological sciences

Ostwald wrote that 'the basic thought upon which this classification rests is that of graded abstraction. . . . All the laws of a former science find application in all the following sciences, since the latter can only contain concepts which by specialization, that is by the addition of further characteristics, have sprung from the concepts of the former or more general sciences.'

CHARACTERISTIC FEATURES OF NINETEENTH-CENTURY CLASSIFICATION

The efforts of classifiers from Francis Bacon onwards were directed towards finding a new rational system of classification to replace that of Aristotle. They took as the 'raw material' of their classifications the traditional fields of study which were current at the time. These fields had already mostly been named and approximately delimited at the beginning of the eighteenth century. The aim the classifiers set themselves was to find a rational grouping or ordering of these disciplines.

This method has been called by Bliss that of the 'scientific and educational consensus'. It assumes, we should note, that the common consciousness of educated society has already divided knowledge into suitably homogeneous blocks. It assumes that the natural historical development of knowledge has already assembled areas of subject-matter with some care and accuracy, and has further performed at least part of the process of carefully dividing these areas into sub-classes.

A second feature of all the classifications so far considered was that the characteristics by which they grouped, divided or arranged the sciences were *subjective*. Bacon's primary division depended on the way the mind approached phenomena—by memory or by reason. All the dichotomies we have examined expressed similar differences in mental approach (speculative and descriptive, pure and empirical, exact and synoptic, and so on). Though Whewell did not rely on such 'faculties of the mind', he also stressed the subjective aspect—'the Ideas which each science involves'.

It must be realised that this was a break with the Aristotelean method. Though Aristotle's primary grouping was subjective (the aim of knowledge), yet he further divided groups of entities, not by means of the mental approach to them, or by means of ideas which they were alleged to embody, but by means of attributes of the entities themselves. The entities studied in his Physics were things that had a separate existence and were changeable. These things were further divided according to their own attributes. For example, Meteorology is the study of changeable things which are inanimate mixtures of the four 'elements'. Chemical substances—Book IV of the *Meteorologica*—are 'perfect' inanimate mixtures, 'perfect' here meaning 'containing some proportion of all four elements'. Al-Farabi further subdivided these as follows:

Perfect inanimate mixtures of elements
- With dissimilar parts
- With similar parts
  - The parts consisting of dissimilar parts, e.g. meat and bone
  - These parts consisting of similar parts, e.g. minerals and stones
  - These parts not further partitioned, e.g. salt, gold, silver

The point under consideration is not whether such a classification corresponds to modern ideas, or even whether it is intelligible to the modern mind. The point is that it is an attempt to divide up the sciences according to the attributes of the material entities studied. It is this objective basis which the nineteenth-century classifiers so far discussed abandoned in favour of subjective, mental, idealist distinctions.

It should also be noted that the subjective mode of dividing the sciences is closely linked with the choice of traditional disciplines as the raw material of classification. The dichotomies propounded by the classifiers gave no help in deciding to which science a particular specific subject belonged. Consider, for example, the study of the flame spectra of substances. Is it Physics or Chemistry? Is it 'practical', as Wilson called Physics, or 'exact theoretical', as he called Chemistry? Is it 'general' or 'applied', to use Goblot's distinction? Could we class phenomena as physical simply by knowing that Schopenhauer considered them to be 'general causal empirical', Bain 'fundamental and abstract', Wundt 'special empirical', and Pearson 'exact concrete'? It is clear that these terms gave no explanation or guidance as to the content of Physics or Chemistry. The contents of these subjects were simply taken to be the traditional materials so assembled, and all that the classifiers did was to give the material a new label.

The third feature of the classifications so far described was that they only carried their divisions a few steps down. The detailed subdivision of the sciences was not attempted, and the dichotomies offered no guidance at all as to the principles by which such subdivision should be carried out.

LEVELS OF AGGREGATION

Although the subjective, idealist approach dominated nineteenth-century classification—and is still with us today—we can trace the emergence of an objective, materialist analysis. As long ago as 1877, the scientist Kekulé suggested the grounds upon which to differentiate the physical sciences. 'Chemistry', he wrote, 'is the science of the statics and dynamics of *atoms:* Physics of *molecules:* whilst Mechanics considers the *masses* of matter consisting of a large number of molecules.'

In opposition to Whewell's views, C. W. Shields stated that 'a philosophical scheme of the sciences should be based upon the facts which support them, rather than upon the ideas which they involve. Such a scheme should fully reflect all the distinct classes of facts which have been scientifically ascertained, and should exhibit all classes of facts in their actual connections.' Shields recognised three 'classes of fact' in the natural sciences—the physical, the chemical and the organical.

In 1884 H. M. Stanley specified the 'classes of fact' more closely as 'orders of aggregation: atoms into molecules, molecules into masses, cell-masses into organisms, and these into societies. Nature is thus a combination of wheels within wheels.' Stanley, like Kekulé, thus defined Chemistry as the science of atoms, Molecular Physics as the science of molecules, Molar Physics as the science of masses and Biology as the science of aggregated cell-masses. 'This classification,' he said, 'presents the general order of the dependence of the sciences. . . . It is also the order of increasing complexity . . . of increasing speciality and concreteness.'

During the 1870s, Frederick Engels had already, in an unpublished manuscript, expressed the issue more exactly from the materialist point of view. In his notes for the *Dialectics of Nature* he thus summarised Hegel's division of the sciences:

Mathematics
Mechanics (the movement of masses)
- Stellar
- Terrestrial

Movement of molecules and atoms
- Molecules (physics)
- Atoms (chemistry)

Organics (the movement of organisms)
- Plants
- Animals

For Engels, as for Aristotle, movement here means activity, change in general. The sciences are thus based on the various material levels of aggregation whose movement is investigated.

The views of Shields, Stanley and Engels were reflected in some of the detailed sub-divisions of classifications whose authors professed other principles. For example, Bain in 1870 distinguished between Molar Physics (Mechanics) and Molecular Physics, while Pearson in 1892 gave an even more 'materialist' breakdown of 'physics' into the four classes: (1) molar (Mechanics, Astronomy); (2) molecular (Crystallography, Hydromechanics, Aeromechanics); (3) atomic (Chemistry, Spectra); and (4) radiation (Light, Heat, Electricity, Magnetism).

The concept of levels of aggregation is now familiar in science—particularly among biologists. Joseph Needham, for example, has expressed it as follows. 'The organic conception of the world involves succession in time and envelopes in space. Taking the latter first, it is obvious that the different levels of organization occur one within the other. Ultimate particles, the proton, electron, etc., build up atoms, atoms build molecules, molecules build large colloidal particles and cell-constituents and paracrystalline phases and the like, these in turn are organized into the living cell.'

The classification of science, the exposition of relations between the specific scientific subjects in a systematic arrangement, must be founded on the material relations of phenomena. The sciences can thus be differentiated by the fact that each treats of a different level of organisation (the nucleus, the atom, the molecule, the cell, the organism, the community, etc.), or a different form of motion (electrical, thermal, optical, etc.). The serial order of the sciences is the arrangement of these levels and forms of motion in their inherent sequence.

From the viewpoint of classification, the significant feature of the concepts developed by Kekulé, Shields, Stanley and Engels is that—unlike the subjective dichotomies discussed earlier—they did provide an objective basis for assembling material into subject groups. We will group as Physics, they suggested, all phenomena which can be described as 'the movement of molecules'. This was not a phrase simply used to label the traditional content of Physics. On the contrary, it was a *definition* of the term 'Physics', and any part of traditional Physics which was not describable as 'the movement of molecules' was to be excluded from the field of Physics.

The materialist analysis made no assumption that history had already assembled homogeneous main classes. It set up its own characteristics of division, defining concretely natural groups of phenomena. These characteristics were a positive guide to the placing of specific subjects. Consider the example of flame spectra discussed earlier. Is it due to the motion and interaction of molecules? No, it is due to the activity of an atom—an internal change in the atom resulting in the external emission of light energy. It is therefore rather to be classed with 'the movement of atoms'—a grouping made by Pearson, as seen above.

The materialists considered here went no further than the subjective classifiers in formulating principles for the further sub-division of the sciences. They suggested a primary division according to levels of material structure and forms of energy, and specified the content of each class as the forms of 'movement' (in the widest sense) found therein. But they did not classify the movements themselves. This work remained to be done.

TWENTIETH-CENTURY CLASSIFICATION

The classifications of science developed by librarians in the late nineteenth and twentieth centuries have been notable for the many sub-divisions they have introduced. Much study of the fine detail of scientific subjects has been carried out by classifiers such as Dewey, Cutter, Brown, Bliss and the Library of Congress. Since I am not attempting here a proper history of classification, all this valuable work cannot be discussed in detail. I can only consider a few of the theoretical developments of the period.

Despite the wealth of new material and ideas which have been introduced into classification, there has remained a good deal of unthinking repetition of the subjective nineteenth-century views already discussed.

For example, the introduction to the abridged English edition of the Universal Decimal Classification notes that it 'does not pretend to provide a philosophical classification of knowledge . . . Nevertheless . . . it is of interest to know how Dewey built up his (original decimal) system. . . . Dewey's starting-point was critical speculative thought, so that group 1 was Philosophy. The individual mind, itself not yet engrossed in the multiplicity of phenomena but concentrated upon the all-embracing, comes next to Religion. . . . Consequently in the first two groups, we find, as it were, a transition from the individual conception to the collective conception, a transition which assumes a more concrete form in group 3, where the Social Sciences are found. . . . Next the community which collects and builds up its knowledge makes use of Language. . . . Then follows first Pure Science. . . . leading to the Applied Sciences. . . .' Dewey's approach was thus just as subjective as that of other nineteenth-century idealists, using dichotomies such as all-embracing and multiple, individual and collective.

The characteristic feature of much of twentieth-century classification, however, has been its empiricism. Theory has been discounted. 'A classification may satisfy in principle a certain philosophical concept and be in accordance with some philosophical system', writes Donker Duyvis, 'but the practical user wants above all to be able to file his documents and papers in an efficient way.' This attitude is perhaps a healthy reaction against the inadequacy of subjective theories, but it is itself an admission of failure.

We cannot avoid theory. Every scientific activity has its theoretical basis. Even the most empirical fact-searching must be guided by an idea of what one is looking for. Many natural scientists may imagine they operate only with undeniable facts, and think that thereby they free themselves from philosophical theories. But, as

R. Flint put it, 'scientific thought is not necessarily self-criticising thought; on the contrary, mere scientific thought, however rigid and methodical, is essentially dogmatic thought in that it rests on untested and uncriticized assumptions. It is reasoned, yet unreflective.' We cannot remain content with such lack of reflection. A rejection of subjective theories should not lead us to abandon the search for a rational basis for the classification of science.

This viewpoint has been stressed by H. E. Bliss, who, however, based his theory on that of Ostwald. Bliss tried to give the 'serial order of sciences' a rational justification by means of an extension of the general–special relation. He listed sciences in the order Logic, Mathematics, Physics, Chemistry, Mineralogy, Astronomy, Geology, Geography, Natural History, Biology, Palaeontology, Botany, Bacteriology, Zoology, etc. Each was a species of the genus 'Science'. But Bliss also considered that the general–special relation was expressed between, say, Physics and Chemistry. Later members of the series he regarded as more 'special' than earlier.

THE SCIENTIFIC AND EDUCATIONAL CONSENSUS

Bliss continued to use the traditional disciplines as the raw material of classification, and claimed that this was justified. 'The fundamental sciences themselves may develop, but in their central concepts, interests, and problems they do not change so radically as to assume new relations, acquire new positions, and require new arrangements. The sciences have remained in the same order for nearly a century, and this order seems likely to persist for centuries to come; for it is based on the order of nature.' This is what Bliss called the scientific and educational consensus, upon which classification should be based.

Let us consider the consequences of this view. If the basic structure of knowledge is relatively permanent, then new facts should be accommodated in it fairly easily—rarely involving new relations, positions or arrangements. If the traditional disciplines represent this permanent basic structure, then new detailed knowledge should be fairly readily assigned to its correct place. Since both the classification and scientific knowledge will be based upon the same broad framework, there should be no conflict between them. Bliss himself believed these consequences to be true. He denied that there was any conflict between logical, scientific classification and practical, bibliographic classification. 'We should join these two hands; the two purposes should be combined. To make the classification conform to the scientific and educational consensus is to make it the more practical.'

I fully agree with Bliss that the more closely a classification reflects the known order of nature, the more practical it is. What I wish to query, however, is whether his conception of the 'scientific and educational consensus' is an accurate enough reflection of the order of nature.

A scientific classification is one which reflects as accurately as possible the relations known to exist in its subject. If the 'consensus', the traditional disciplines, are scientific in this sense, why do we find a conflict between scientific classification and book classification? Why do the subjects of books sometimes appear to fit so badly into the latest scientific system? Why do we find that the idea of a scientific classification is vigorously opposed by classifiers such as E. Wyndham Hulme, who pin their faith on 'literary warrant'?

Over 60 years ago, Hulme maintained that the functions of 'philosophical' and book classifications were distinct: the latter maps literature, not science. The division and coordination of classes in literature, he declared, 'is determined mainly upon formal and non-philosophic lines'. Book classification was 'the plotting of areas pre-existing in literature', and coincidence with a philosophical order was no guarantee of accuracy. The warrant for coordination was purely literary. 'The appeal is not to reason, but to facts revealed by the literary census.' As an example, Hulme cited the strong literary warrant for coordinating Food and Drugs, under Chemical Analysis, rather than Food and Drink. Since 'separate provision must be made for every distinct book class', and book classes do not correspond to the order of nature, 'the result is a welter of cross classifications and of overlapping areas of definition, for the reception of which the framework of philosophic classification is quite insufficient'.

Why do book classes fail to correspond to 'philosophical' classes? Hulme indicated a reason in the following remarks. Books continue to be written, he said, on such subjects as Metals and Metalloids. Yet these terms, 'while retained for convenience, are *no longer* regarded as conveying distinctions of real scientific value. In Inorganic Chemistry, what has philosophy to offer? Merely a classification by the names of the elements for which practically no literature *in book form* exists. No monograph, for instance, has yet been published on the Chemistry of Iron or of Gold. . . . (If such terms are provided in the schedule) the field will be found occupied by the *snippet* literature with definitions of details for which no literature is likely to be forthcoming.' (My emphases throughout, B.C.V.)

This passage throws light on the conflict between 'literary warrant' and 'logical classification'. It shows that the real conflict is between old and new concepts in the subject classified. The concept of Metalloid in Chemistry was 'philosophical' enough once, but the Periodic Classification made it outmoded: yet it persists in book titles, and so must be provided for in a book classification. On the other hand, the newer concepts must also be provided for in the classification. Today, 60 years after Hulme wrote, we have many monographs on individual elements—even on individual compounds. We can never safely prophesy of a subject, 'no literature is likely to be forthcoming'. On the contrary, almost every 'snippet' subject eventually develops its own textbooks.

In fact, the reason why the subject of books sometimes fits badly into an up-to-date scientific classification scheme is that the *book* literature of a subject tends to lag behind the detailed development of that subject. The *article* literature, expressing the newest thought on the subject, corresponds much more closely to a scientific classification—there were already innumerable *articles* on the Chemistry of Iron when Hulme wrote in 1911. Specialist monographs are also relatively closely based on the latest scientific classification. It is the more extensive subject headings which exhibit this time-lag most markedly, and are most likely to be based on an earlier, outmoded classification. This classification is now 'unscientific', retained only as a tradition, a conventional division of the subject.

This point has been made by recent writers on Physics, which is still conventionally divided into Mechanics, Acoustics, Light, and so on. Otto Blüh notes that 'the classification of physics originally followed the types of sense perception. . . . But this method of division has lost its meaning.' The same point is emphasised by Professor Herbert Dingle, in a discussion of the problem of comparing the science of today with that of 100 years ago. 'How shall we sub-divide the field to be covered? The classification most suitable for 1851 becomes absurd for 1951; that for 1951 is impossible for 1851. In the earlier years physics fell naturally into the conventional groups—mechanics, heat, light and so on; these were separate sciences, in very different stages of development, each following its own line of growth independently of the others. Today the boundaries between them have almost completely gone. . . . Biology tells the same tale. . . . The general picture is one of extension, unification and generalization.'

Science, in fact, has outgrown the traditional disciplines of Physics, yet they linger on in book titles and educational curricula. This time-lag, this slow abandonment of outworn ideas, was taken by Bliss to be due to the 'relative permanence' of these ideas. In reality, it is due much more to the 'relative inertia' of writers and teachers—that is, inertia relative to the dynamic advance of science itself. Undoubtedly, the continued use of conventional titles in literature creates a problem for classification. As Richardson emphasised, 'the whole of things includes not only things present, but things past', and we must find ways of including both in our classification, but not by assigning to past ideas a validity and a vitality which they do not possess, distorting their relations to present ideas.

The 'scientific and educational consensus' based on traditional disciplines is an idealist conception, as Broadfield has already stressed. Bliss laid down that the verification of knowledge depends 'intellectually on a consensus of minds accepting these truths'. Broadfield replies that this suggests 'that the truth of what is verified depends on agreement between minds who consent to its being true'. But the real relations of science are not determined by the ways scientists think things to be related, but 'by the nature and relations of their knowable concepts'—i.e. by the nature of the external reality which the scientists explore.

THE ORDER OF THINGS

The views of the nineteenth-century materialists have not been without influence on twentieth-century library classification—the writings of Shields, in particular, were well known. Thus we find Cutter remarking that 'science proceeds from the molecular to the molar'. But the most extended discussion is that of E. C. Richardson.

In his lectures (1901) Richardson started from the statement that 'classification is the putting together of like things. . . . A *thing* as subject for classification is whatever has separate existence. Whether its substance is matter or motion or spirit is indifferent. . . . Things include the things which make man (nature and environment) and the things which man makes. . . . Moreover, ideas are also real things, calling for a place in the order of things. . . . Things, therefore, include nature, ideas and art. . . . *The order of the sciences is the order of things.*'

On this basis, he went on to distinguish various levels of complexity among natural things. He started with 'ions' (we would today call them 'sub-atomic particles'). These unite to form atoms, each of which is itself 'a discrete thing, which moves as a whole, so that its parts do not part company under ordinary conditions'. These atoms 'themselves unite with one another in various degrees of complexity to form another series of things, each of which moves about as a whole'—molecules. Groups of molecules are formed into masses, and these in turn into planets. 'This same process continues among complex molecules . . . which become organized into independent groups which we call cells.' These unite into organisms. Among human organisms, 'there is again a tendency towards union into . . . societies'.

This, then, according to Richardson, was the order of things upon which the order of the sciences was to be based. 'The science itself is nothing apart from the things or facts with which it deals. To define a science one must define the group of facts with which it deals.' He was convinced that the notion that classification is one of subjects rather than of objects was 'a profound theoretical and practical mistake, leading to endless confusion'.

'It is hardly too much to say that the chief drawback to progress in the conception of the whole of things has been this very confusion between the inner realm of ideas and the outer realm of things. Men have somehow looked on their own ideas as the place where the order of things was to be discovered, and many systems of classification, e.g. that of Bacon, have been founded on subjective laws.' Richardson went on to criticise 'the vexatious subjective distinctions of abstract and concrete, pure and applied'. 'It is, of course', he continued, 'within the mind's sphere of ideas that a man must work. He does not deal with the things themselves directly so much as with little brain maps of them. Yet at the same time it is true that these very brain maps themselves have no meaning or value save as they are true and complete images of the outer reality. It is the outer universe which is the starting point and fixed factor of all the search for order. As the inner idea corresponds with the real fact or truth, or not, it is worthy or worthless.'

In one way, any flat contrast between subjective and objective classification is a false one. It is true that for a classification to be valid in practice, the characteristics it uses must correspond with reality. But the entities we classify have innumerable real characteristics. Which of them we choose in constructing a particular classification will depend, as Broadfield stressed, on the *purpose* of the scheme. Whereas the choice of characteristics is subjective, a matter for decision by the classificationist, the characteristics chosen should be objectively found in nature. The great mass of factual detail which has been crammed into the schedules of modern classification schemes must not blind us to the fact that their structure is all too often based on subjective principles of division. The lesson of this historical survey was firmly emphasised by

Richardson himself: 'The starting point of all progress towards getting a clear conception of the order of the sciences is the axiom: the order of the sciences is the order of things.'

ANALYTICO-SYNTHETIC CLASSIFICATION

The major theoretical developments of twentieth-century classification have been most clearly formulated by S. R. Ranganathan, who has, however, concentrated his attention in an almost mathematical way on the general structure of knowledge.

He pointed out that the field of knowledge, i.e. the number of specific subjects which have to be related in a classification, is potentially infinite: between any two points in the system an indefinite number of new points might have to be inserted. Moreover, knowledge is multidimensional: the interconnections of each concept spread out in many directions, and usually each subject is a synthesis of several such multiply connected concepts. Subjects so formed are said to show 'lamination'. A still looser form of combination, 'assemblage', exists when two 'universes of discourse, each infinite and self-contained in some definable sense, may have to be considered in relation to one another'.

Puranik, a colleague of Ranganathan, analysed the subject matter of five different scholarly journals over about 50 years, and showed that while in 1900 over half of the sample subjects were 'simple' concepts, in 1950 over 85% were compounds exhibiting lamination or assemblage. It is with such concrete evidence as this in mind that Ranganathan designed a classification in which compound subjects were built up synthetically from 'elementary' concepts. Some of his most original work has been the formulation of notational devices to represent such combinations by a system of ordinal numbers.

In constructing his Colon Classification, Ranganathan started with a 'universe' of entities (concrete and conceptual existents). Each entity has many attributes, and some of these can serve as characteristics for the division of the universe into groups or classes. The attributes usable as characteristics, he stated, should be relevant, ascertainable and relatively permanent. The classes so derived can then be ranked to give a helpful order in array, according to a number of postulated principles: decreasing extension, increasing concreteness, evolutionary, chronological, geographical, increasing complexity and canonical (conventional). Ranganathan further formulated eight sets of rather vaguely related 'seminal' concepts, which he ranked in array in the following order:

1. Fundamentals, preliminary, person, nomenclature.
2. Structure, morphology, parts, form, constitution, sources.
2. Function, physiology, activity, syntax, method, analysis, technique, transformation, distribution, interpretation.
4. Fault, pathology, social ills, disease, transport, interlinking, synthesis, hybrid.
5. Fluidity, liquid, sap, blood, ocean, control, beauty, energy, light, organic, environment.
6. Correction, remedy, financing, money, mysticism, abnormality, phylogeny, evolution, relief, movement.
7. Personality, magnetism, ontogeny, composition, growth, produce, value.
8. Management, organisation, testing, equipment, fitness.

Each class so ranked is itself a fresh 'universe' susceptible of division by another characteristic. Successive division by a train of characteristics gives rise to a chain of classes, which should be appropriately modulated. The meaning of 'modulation', though clear enough to the intuition, is not easy to define.

In the practice of classification the original universe to be divided was one of the main classes, defined by Ranganathan as 'the fairly homogeneous, conventional regions of knowledge, which together form the first order array of classes which are mutually exclusive and totally exhaustive of the field of knowledge'.

He later recognised other 'original universes' or 'basic subjects'. As summarised by Neelameghan, traditional main classes might give rise to new basic subjects in four ways:

(1) Distillation—e.g., from such topics as 'research methods in physics', 'research methods in chemistry', 'research methods in biology', is distilled out a new basic subject, Research Methodology.
(2) Specialisation—e.g., the development as adjuncts to Biology of the specialisations Cytology, Microbiology, Molecular Biology.
(3) Fusion—e.g. the formation of interdisciplinary subjects such as Astrophysics, Biochemistry, Geomechanics.
(4) Clustering—e.g. Ocean Sciences, Materials Sciences.

Further, each of the primary basic subjects so formed might give rise to secondary basic subjects that could also be treated as 'original universes', in the following ways:

(1) Fission—e.g. Physics is canonically divided into the basic subjects Sound, Heat, Radiation, etc.
(2) Lamination—a primary basic subject expounded according to a particular system of thought (e.g. Relativity Physics), or within special boundary conditions (e.g. Child Medicine), or in a particular environment (e.g. Freshwater Biology).

Ranganathan next suggested that any such original universe could be divided by more than one train of characteristics, each train giving rise to one facet of the universe. The facets encountered in the field of knowledge were all manifestations of five fundamental *categories*—Time, Space, Energy, Matter and Personality—whose nature can be seen from the following examples.

Time and *space* were relatively simple categories. *Energy* covered facets which Ranganathan had previously named problem, method, process, operation, handling and technique. *Matter* comprised constituent materials of all kinds, and *personalities* included libraries, numbers, equations, wavelengths of radiation, engineering works, chemical substances, organisms and organs, crops, manures, religions, art styles, literary products, languages, social groups, communities and states. In recent years Ranganathan modified his views of the *matter* category, and included within it not only materials but also properties and methods, excluding these from the *energy* category.

Ranganathan further suggested that the categories of *energy*, *matter* and *personality* could be manifested more than once within a given original universe. The first *energy* facet could introduce a second 'round' of the *personality* and *matter* categories, the second *energy* facet which follows could introduce a third round, and so on. *Time* and *space* could figure only in the last round.

Even within any one facet, further *levels* of complexity were evident. The *time* facet was dissected unidimensionally into a series of conventional time intervals, but there was sometimes need to indicate a stretch of time which overlapped these conventional divisions. The *space* facet was dissected two-dimensionally into conventional geographical areas, but other groupings must be allowed for, in which the areas grouped were not necessarily adjacent (disjunctive incidence): for example (a) countries grouped round a certain sea, (b) climatic zones, (c) areas with common social, industrial, religious, etc., characteristics, (d) empires, (e) international associations. Physical features also cut across the conventional groupings. The *energy* category, Ranganathan postulated, was subject to no such complexity. He suggested that it could manifest itself only as a single array of terms in any one facet. *Personality*, however, exhibited levels—the whole personality might be sub-divided into parts or organs, these in turn into sub-parts, and so on.

The above description of the structure of knowledge is static. But Ranganathan stressed that the universe of knowledge was not static: 'It is a dynamic continuum. Its structure is therefore ever changing. Hence it is not sufficient to know its structure at the present moment.

We should also understand its mode of development.' He indicated five modes of formation of new specific subjects: (a) by the lamination of terms from different facets within the same original universe, and (b) by the loose assemblage of terms from different original universes; new terms were introduced into the structure by (c) the denudation of a class to form a sub-class; by (d) the dissection of a class to form an array of sub-classes; or by (e) the interpolation of a new term in an existing array. This kind of development went on continually. As such changes accumulated in a particular main class, it became necessary to reorganise the ranking of terms in particular arrays, or even in a whole facet. Further, it became necessary to reshape the facet formula of the class. Ranganathan conjectured that such adjustments might occur once in a generation. At one time he believed that it was less frequently necessary to establish a new main class, but recently he greatly increased the number of 'basic subjects'.

Many of these proposals came as a fresh wind, freeing classification from the rigidity of thought with which it had been afflicted, pointing towards the formation of a new, flexible instrument which could incorporate the complex relations of the phenomena to be classified. Nevertheless, his ideas still need to be subject to critical examination—in particular, his rather restricting use of 'fundamental categories', and the methods proposed for developing new basic subjects.

THE UNIVERSAL DECIMAL CLASSIFICATION

Although the theory of analytico-synthetic classification has been most extensively discussed by Ranganathan, the earliest general use of the technique—and still in practice the most important example—is the Universal Decimal Classification of P. Otlet and H. La Fontaine. The main class structure of the classification is that of Dewey, but in its detail and its use of synthesis it has moved a long way from the original Decimal Classification.

The editors of UDC stress its practical, empirical character. It is allegedly not based on any philosophical conception of the structure of knowledge. It is simply a system of practical devices for 'multi-dimensional analysis'. These devices have been summarised by Donker Duyvis as follows:

Degree of complexity
- Multiplicity: composite subjects can be represented by linking their class numbers by / and +
- Individuality: further sub-divisions of a class can be labelled A to Z.

Relationship
 General: symbolised by linking related class numbers by the colon.
 Points of view: special numbers introduced by the double zero.
Special common features in a basic class: designated by numbers introduced by the dash or zero.
Space, time, race, language and bibliographical form schedules are provided.

Examples of the combinatory devices mentioned by Donker Duyvis are given below:

Form	
53(032)	Physics dictionary
624.131.439.4(043)	Thesis on the compressive strength of soil
593.171.4(086.3)	Microscopic preparation of *Paramecium*
637.2(088.7)	Trade marks for bstter
622.344.5(091)	History of lead mines
Language	
61(03)=40	Medical encyclopedia in French
413.2=3=82=84	German–Russian–Polish dictionary
Race	
159.9(=8)	Psychology of Slavs
299.6(=936.532)	Religion of the Hereros (in Africa)
Place	
639.2(261.2)	Fisheries in the North Sea
622.344.5(43-15)	Lead mines in Western Germany
Time	
355.49(384) '-0480'	Naval battle at Salamis in 480 B.C.
523.78 '1955.06.30.12.25'	Eclipse of the sun at 12.25 on 30 June, 1955
Point of view	
629.001.33	Classification of ships
621.833.002.3	Materials for gears
666.15.004.55	Cleaning of sheet glass
The colon	
621.94: 658.58	Lathe maintenance
536.33: 535.34	Absorption of heat radiation
549.646.3: 531.423	Specific weight of aquamarine
The dash	
534.37–8	Damping of ultrasonic vibrations
535.217–3	Chemical effect of ultraviolet light
541.13–38	Electrochemistry of salts
611.61–018	Histology of the kidneys
616.61–004	Hardening of the kidneys
629.114.4–83	Electric-driven vans
The zero	
576.852.093	Culture methods for actinomycetes
621.365.2.078	Automatic regulation of arc furnaces
629.124.2.014.6	Rudders of tugboats
636.15.083.7	Harness for cart-horses
661.721.094.4	Halogenation of methanol
The apostrophe (restricted to 546, 547, 678)	
645.623'32'226	Potassium aluminium sulphate

Though the UDC claims to be no more than a system of practical devices, and though its general structure, based on Dewey, is often defective, its detailed schedules—especially in Technology—reveal the facet pattern and display a great volume of valuable analysis upon which future classifications—both general and special—will be bound to draw. The relation of both UDC and Colon to the classificatory categories developed in this book is examined in Appendix C.

CONCLUSIONS

This brief historical survey reveals the main trends in the classification of science. We note first the ever-widening extent of the field of knowledge embraced by the term 'Science—pure and applied'. Additions to the Aristotelean *physica* have been continual—Algebra from the Arabs, 'mechanical arts' in the Middle Ages and later, numerous new scientific disciplines in the nineteenth and twentieth centuries. At the same time as the field has widened, it has grown in complexity, and classifications have had to cope with an increasing amount of detail. As Ranganathan summed it up, the field of knowledge is potentially infinite, and ever-growing.

In classifying this knowledge, the old subjective modes of division have been increasingly—though not yet entirely—abandoned. The attempts to divide science into subjectively defined disciplines have proved unsuccessful. A more objective approach, based on the concrete and specific entities of science, has now been developed. The restriction of classification to the traditional tree-of-knowledge has given way to the recognition of categories which introduce relations other than the generic. The field of knowledge is multi-dimensional; terms and concepts are multiply connected.

The full implications of these developments for the classification of science have not yet been worked out. It may fairly be claimed that there has now been evolved a technique—that of facet analysis—by which a special, relatively narrow field of knowledge can be systematised. Some may think that this is all that is required. Because of the defects of general classifications such as those mentioned in this appendix, there has been a certain reaction against such schemes, and more stress is laid on the provision of special classifications. However, the need for a wider, universal classification scheme—at least for the whole of science and technology—is still felt. Not only is it needed for use in the more comprehensive libraries and bibliographies, but also it would be of great help in the construction of special schemes. Many of these use very similar categories—e.g. for physical properties, for machine parts, for chemical processes—and a comprehensive enumeration of such categories in a general scheme would provide a source on which special schemes could draw, thus saving much duplication of effort.

The exact form of a general classification of science, in line with modern developments, has yet to be settled. We have the clues provided by such writers as Richardson and Ranganathan, much detail worked out by the makers of the UDC. In Chapters 2 and 3 there are a few suggestions as to the relation of basic subject fields and traditional disciplines. But many problems remain. While the construction of a special scheme is the immediate task facing many a librarian, or the use of such a scheme in a mechanised search system, the long-term work of providing a new general classification of science and technology must also be faced.

SOME REPRESENTATIVE CLASSIFICATIONS OF SCIENCE

Nasir-al-Din (1250)
Speculative sciences
- Metaphysics
- Mathematics
  - Elements
    - Geometry
    - Arithmetic
    - Astronomy
    - Music
  - Derivatives
    - Perspective
    - Optics
    - Algebra
    - Mechanics

Physics
- Elements
  - Motion in general
  - Motion of stars
  - Coming to be and passing away
  - Meteorology
  - Mineralogy
  - Plants
  - Animals
- Derivatives
  - Medicine
  - Astrology
  - Agriculture

Hugh of St. Victor (1140)
Theoretical sciences
- Mathematics
  - Arithmetic
  - Geometry
  - Music
  - Astronomy
- Physics, including
  - Meteorology
  - Mineralogy
  - Plants
  - Animals

Mechanical arts
- Cloth manufacture
- Armour-making
- Navigation
- Agriculture
- Hunting
- Medicine
- Theatrical science

Roger Bacon (1250)
Mathematics
- Arithmetic
- Geometry
- Astronomy
- Music

Physics
- First principles
- Rest and motion
- The four elements
- Compounds
  - Minerals
  - Plants
  - Animals
- Coming to be and passing away
- Change of quality
- Growth and dimination

Special science
- Perspective
- Astrology
- Statics
- Alchemy
- Agriculture
- Medicine

Francis Bacon (1605)
Natural history
- Generations
  - Astronomy, descriptive
  - Meteorology
  - Earth and sea
  - Four elements
  - Plants and animals
- Marvels
- Arts
  - Agriculture
  - Mechanical
  - etc.

Natural philosophy

Francis Bacon (1605) (*continued*)

- Physics (principles)
- Metaphysics
  - Mathematics
    - Pure
      - Geometry
      - Arithmetic
    - Mixed
      - Perspective
      - Music
      - Astronomy, theoretical
      - Architecture
      - Engineering

Gesner (1548)

- Preparative
  - Necessary
    - Linguistic
      - Grammar
      - Dialectics
      - Rhetoric
      - Poetics
    - Mathematics
      - Arithmetic
      - Geometry, Optics
      - Music
      - Astronomy
      - Astrology
  - Ornamental
    - Divination
    - Geography
    - History
    - Diverse arts
- Substantial
  - Natural philosophy
  - Metaphysics
  - Morals
  - Economics
  - Politics
  - Law
  - Medicine
  - Christian theology

Royal Society (1725)

- Mathematics
- Surveying
- Optics
- Astronomy
- Mechanics, acoustics
- Hydrostatics, hydraulics
- Geography, navigation
- Architecture, ship-building
- Perspective
- Music
- Physiology, meteorology, pneumatics
- Hydrology
- Mineralogy
- Magnetics
- Botany, agriculture
- Zoology
- Anatomy, physic
- Pharmacy, chemistry

Young (1807)

- Mathematics
  - Arithmetic, algebra
  - Geometry
  - Comparison of variables
  - Curves
- Natural philosophy and mechanical arts
  - Mechanics, including Practical mechanics
  - Hydrodynamics
  - Acoustics
  - Optics
  - Physics
    - Astronomy
    - Properties of matter
    - Heat and cold
    - Electricity
    - Magnetism
    - Meteorology
- Natural history
  - Mineralogy
  - Botany
  - Zoology
  - Agriculture

Arnott (1827)

- Mind, including logic and mathematics
- Life
  - Vegetable (plants and agriculture)
  - Animal, including medicine
- Chemistry, including mineralogy, geology, pharmacy
- Physics
  - Mechanics
  - Hydromechanics
    - Hydrostatics
    - Pneumatics
    - Hydraulics
    - Acoustics
  - Heat
  - Light
  - Electricity
  - Magnetism
  - Astronomy

Lindsay (1845)

- Science
  - Speculative
    - Grammar
    - Mathematics
    - Mechanics
    - Astronomy
    - Acoustics
    - Optics
    - Electricity
    - Magnetism
    - Chemistry
    - Geology
    - Geography
    - Botany
    - Zoology
    - Comparative anatomy
    - Biology
    - Noology
    - Psychology
  - Practical
    - Hunting
    - Agriculture
    - Domestic economy
    - Mechanical arts
    - Medicine
    - War
    - Navigation

Wilson (1856)

- Theoretical
  - Exact
    - Meteorology
    - Uranography
    - Geology
    - Geography
    - Chemistry

Wilson (1856) (*continued*)

-
    - Mineralogy
    - Anatomy
    - Physiology
    - Botany
    - Zoology
    - Ethnology
    - Psychology
    - History
  - Pure
    - Arithmetic
    - Geometry
    - Algebra
    - Calculus
- Practical
  - Mechanics
  - Astronomy
  - Hydromechanics
  - Pneumatics
  - Acoustics
  - Optics
- Productive
  - Agriculture
  - Metallurgy
  - Technology
  - Typography
  - Medicine

Whewell (1840)

- Pure mathematical
  - Geometry
  - Arithmetic
  - Algebra
  - Differentials
- Pure motional
  - Pure mechanism
  - Formal astronomy
- Mechanical
  - Statics
  - Dynamics
  - Hydrostatics
  - Hydrodynamics
  - Physical astronomy
- Secondary mechanical
  - Acoustics
  - Formal optics
  - Physical optics
  - Thermotics
  - Atmology
- Analytico-mechanical
  - Electricity
  - Magnetism
- Analytical

Whewell (1840) (*continued*)

-
  - Chemistry
- Analytico-classificatory
  - Crystallography
  - Mineralogy
- Classificatory
  - Botany
  - Zoology
  - Comparative anatomy
- Organical
  - Biology
- Palaeotiological
  - Geology
  - Distribution of living organisms

Dove (1850)

- Mathematical
  - Logic
  - Arithmetic
  - Algebra
  - Geometry
  - Statistics
- Physical
  - Dynamics
  - Mechanics
  - Acoustics
  - Optics
  - Thermology
  - Magnetism
  - Electricity
  - Chemistry
- Organisational
  - Vegetable and Animal physiology

Schopenhauer (1800)

- Pure science
  - Being
    - Spatial
      - Geometry
    - Temporal
      - Arithmetic
      - Algebra
  - Knowledge
    - Logic
- Empirical
  - Causes
    - General
      - Mechanics
      - Hydromechanics
      - Physics
      - Chemistry

Schopenhauer (1800) (*continued*)

-
  -
    - Special
      - Astronomy
      - Mineralogy
      - Geology
      - Technology
      - Pharmacy
  - Stimuli
    - General
      - Physiology
      - Anatomy
    - Special
      - Botany
      - Zoology
      - Pathology
      - Therapeutics

Wundt (1889)

- Formal sciences
  - General
    - Quantitative
      - Algebra
      - Functions
    - Qualitative
      - Groups
  - Special
    - Number
      - Arithmetic
      - Theory
    - Space
      - Geometry
    - Motions
      - Kinematics
- Empirical sciences
  - Natural processes
    - General
      - Dynamics
    - Special
      - Physics
      - Chemistry
  - Natural objects
    - Astronomy
    - Geography
    - Natural history
      - Mineralogy
      - Botany
      - Zoology
  - Natural processes in natural objects
    - Concrete physics and chemistry
      - Astrophysics
      - Geophysics

Wundt (1889) (*continued*)

- Physiology, etc.
- History of development, evolution, cosmology, geology

Bain (1870)

- Fundamental or abstract
  - Logic
  - Mathematics
  - Physics
    - Molar
      - Kinematics
      - Statics
      - Dynamics
    - Molecular
      - Attractions
      - Heat
      - Light
      - Electricity
  - Chemistry
  - Biology
  - Psychology
- Dependent or concrete
  - Applied mechanics
  - Hydromechanics
  - Aerostatics and dynamics
  - Acoustics
  - Astronomy
  - Mineralogy
  - Botany
  - Zoology
  - Meteorology
  - Geography
  - Sociology
  - Philology
- Practical
  - Building
  - Dyeing
  - Navigation
  - Engineering
  - Baking
  - Brewing, etc.

Dewey (1876)

- Science
  - Mathematics
  - Astronomy
  - Physics
  - Chemistry
  - Geology
  - Palaeontology

Dewey (1876) (*continued*)

  - Biology
  - Botany
  - Zoology
- Useful arts
  - Medicine
  - Engineering
  - Agriculture
  - Domest c economy
  - Commun cation and commerce
  - Chemical technology
  - Manufactures
  - Mechanic trades
  - Building

Masaryk (1886)

- Abstract sciences
  - Mathematics
  - Mechanics
  - Physics
  - Chemistry
  - Biology
  - Psychology
  - Sociology
- Concrete
  - Geometry
  - Astronomy
  - Acoustics
  - Hydromechanics
  - Aeromechanics
  - Cosmography
  - Cosmology
  - Botany
  - Zoology
  - etc.
- Practical
  - Calculation
  - Descriptive geometry
  - Industrial arts
  - Technology
  - Hygiene
  - Medicine
  - etc.

Pearson (1892)

- Abstract sciences
  - Qualitative
    - Logic
    - Grammar
  - Quantitative
    - Discrete
      - Arithmetic

Pearson (1892) (*continued*)

      - Algebra
      - etc.
    - Change of quantity
      - Functions
      - Calculus
      - etc.
  - Spatial
    - Geometry
    - Trigonometry
    - Mensuration
    - etc.
  - Temporal
    - Kinematics
    - etc.
- Concrete sciences
  - Inorganic
    - Exact (physics)
      - Molar
        - Mechanics
        - Astronomy
      - Molecular
        - Crystallography
        - Hydro-mechanics
        - Aeromechanics
        - Tides, etc.
      - Atomic
        - Theoretical chemistry
        - Spectrum analysis
        - Astrophysics
      - Radiation
        - Light
        - Heat
        - Electricity
        - Magnetism
        - etc.
    - Synoptic
      - Geology
      - Geography
      - Meteorology
      - Mineralogy
      - Chemistry
      - etc.
  - Organic
    - Spatial
      - Natural history
      - Ecology
      - Geographical distribution
    - Temporal

Pearson (1892)
(*continued*)
- History
  - Phylogeny
  - etc.
- Biology
  - Morphology
  - Physiology

Goblot (1898)
- Mathematical
  - Arithmetic
  - Algebra
  - Geometry
  - Mechanics
- Theoretical
  - Cosmology
    - General
  - Attraction
      - Heat
      - Optics
    - Acoustics
      - Electrology
    - Applied
      - Chemistry and mineralogy
      - Astronomy and physical geography
      - Cosmogony
        - Geology
        - Palaeontology
  - Bio-psycho-sociology
    - General
      - Physiology
      - Anatomy
    - Applied
      - Biology
        - Botany
        - Zoology
        - Anthropology
      - Geography
      - History
- Practical
  - Mechanical arts
  - Medicine, etc.

Library of Congress (1904)
- Science
  - Mathematics
  - Astronomy
  - Physics
  - Chemistry, with crystallography
  - Geology, with mineralogy, palaeontology
  - Natural history
  - Botany
  - Zoology
  - Human anatomy and physiology
  - Bacteriology
- Medicine
- Agriculture
- Technology

Brown (1906)
- Physical science
  - Physics, dynamics
    - Engineering
    - Architecture
    - Transport
  - Electricity
  - Optics
  - Heat
  - Acoustics
  - Astronomy
  - Physiography
    - Hydrography, hydrostatics
    - Meteorology, pneumatics
  - Geology
    - Crystallography
    - Mineralogy
    - Metallurgy
    - Mining
  - Chemistry and technology
- Biological science
  - Biology
  - Botany
  - Zoology
- Ethnology and medicine
- Economic biology and domestic arts

Bliss (1940)
- General science
  - Logic
  - Mathematics
  - Metrology
  - Statistics
- Physics
  - Applied physics
- Chemistry
  - Chemical techology
  - Mineralogy
- Cosmology
  - Astronomy
  - Geology
  - Geography
  - Natural history
  - Microscopy
- Biology
  - Palaeontology
  - Biogeography
- Botany
  - Bacteriology
- Zoology
- Anthropology
  - Medicine
- Useful arts
  - Industrial arts
  - Agriculture
  - Engineering
  - Building
  - etc.

Hulme (1912)
- Physical geography
- Geology
- Crystallography
- Mineralogy
- Petrology
- Stratigraphical geology
- Palaeontology
- Economic geology
- Mineral industries
- Mine engineering
- Coal
- Iron and steel
- Foundry work
- Alloys; metallography
- Non-ferrous metallurgy
- Assaying
- Systematic metallurgy
- Furnaces
- Fuel
- Peat industry
- Destructive distillation
- Artificial lighting
- Mineral oils
- Gas
- Acetylene and carbide
- Oils and fats

Hulme (1912) (*continued*)	Gums, resins	Practical chemistry
Paints, pigments, varnishes	Plastics	Chemical analysis
Waxes, lubricants	Rubber	Inorganic chemistry
Essential oils	Cellulose, paper	Physical chemistry
Soaps, candles	Leather	Organic chemistry
	Chemical technology	Biochemistry

BIBLIOGRAPHY

General: R. FLINT. *Philosophy as Scientia Scientiarum and a History of Classifications of the Science*. London, 1904

W. C. BERWICK SAYERS. *Manual of Classification*. London, 3rd ed., 1955

H. E. BLISS. *The Organization of Knowledge and the System of the Sciences*. New York, 1929 (Historical, pp. 308–414)

J. LUND and M. TAUBE. 'Non-expansive classification system'. *Lib. Q.* 1937, **7,** 373–94

J. MOUHASSEB. *Essai sur la Classification des Sciences*. Damascus, 1953 (Historical survey)

E. C. RICHARDSON. *Classification, Theoretical and Practical*. New York, 3rd ed., 1930 (Includes an historical list)

E. DE GROLIER. *Théorie et pratique des classifications documentaires*. Paris, 1956 (Very helpful historical section)

E. DE GROLIER. 'Quelques travaux récents en matière de classification encyclopédique'. *Bull. Bibl. France*, 1970, **15,** 99–126

J. MILLS. 'Library classification'. *J. Docum.*, 1970, **26,** 120–60

M. F. TAUBER and E. WISE. *Classification Systems*. New Brunswick, 1961.

Plato: W. WHEWELL. *Proc. Cambr. Phil. Soc.*, 1855

Aristotle: W. D. ROSS. *Aristotle*. London, 4th ed., 1945

al-Farabi: M. CLAGETT. *Isis*, 1948, **39,** 29–36

Nasir al-din: J. STEPHENSON. *Isis*, 1923, **5,** 329

Hugh of St. Victor: J. MARIETAN. *Probléme de la Classification des Sciences d'Aristote à St. Thomas*. Paris, 1901

Roger Bacon: R. STEELE. *Studies in the History of Science*, vol. 2, 121–50. Oxford, 1921

CONRAD GESNER. *Pandectarum sive Partitionum Universalium*. Zurich, 1548

FRANCIS BACON. *Advancement of Learning*. 1605

JOHN LOWTHORP. *The Philosophical Transactions and Collections Abridg'd*. London, 1725

THOMAS YOUNG. *Course of Lectures on Natural Philosophy and the Mechanical Arts*, vol. 2. London, 1807

NEIL ARNOTT. *Elements of Physics*. London, 1827

WILLIAM WHEWELL. *Novum Organon Renovatum*. London, 3rd ed., 1858

MELVIL DEWEY. See *Universal Decimal Classification, abridged, English edition* (British Standard 1000 A, London, 1948)

A. KEKULE. *Nature*, 1877, **17,** 55

H. M. STANLEY. *Mind*, April 1884, **34**

F. ENGELS. *Dialectics of Nature.* English translation, London, 1940

C. W. SHIELDS. *Philosophia Ultima.* New York, 3rd ed., 1888

C. A. CUTTER. *Trans. and Proc. of Second Internat. Lib. Conf.* London, 1898

H. E. BLISS. *Bibliographic Classification*, vol. 1. New York, 1940; *The Organization of Knowledge.* New York, 1929

A. BROADFIELD. *The Philosophy of Classification.* London, 1946

H. DINGLE in *A Century of Science*, Chapter XX. London, 1951

E. W. HULME. *Principles of Book Classification.* Reprinted London, 1950

J. D. Brown: J. D. STEWART. *Rev. Docum.*, 1950, **17,** 56–63

S. R. RANGANATHAN. *Colon Classification.* London, 6th ed., 1960; *Philosophy of Library Classification.* Copenhagen, 1951; *Prolegomena to Library Classification.* London, 2nd ed., 1957 and London, 3rd ed., 1967

Universal Decimal Classification: British Standard 1000A, *U.D.C., Abridged English edition.* London, 2nd ed., 1957; F. DONKER DUYVIS, *Rev. Docum.*, 1951, **18,** 99–105; J. M. PERRAULT. *Towards a Theory for UDC.* London, 1969; A. C. FOSKETT. *The UDC.* London, 1973

APPENDIX B

EXAMPLES OF FACETED CLASSIFICATIONS

In this Appendix are presented two examples of faceted classification schedules—one in Soil Science and the other (drawn up by D. J. Foskett) in Container Manufacture.

SOIL SCIENCE

The schedule which follows exhibits the following facets:

9 Kinds of soil (variously subdivided)
8 Structure (including strata, horizons, particles)
7 Constituents (chemical, parent material, organisms)
6 Properties (physical, physico-chemical, biological)
5 Processes in soil
4 Operations on soil (including amendment and amendment materials)
3 Laboratory techniques
1 General

The facet indicators are the numerals listed above. Terms in each facet are symbolised by small letters (excluding l and o). Two other symbols are used. The oblique stroke represents 'relation'—particularly 'influence' or 'effect'. Thus 6s/4f is 'salinity in relation to irrigation'. The relationship can be between two facets, as here, or within a facet, as in 5ib/j, 'the effect of evaporation on infiltration'. The other symbol used is the hyphen, which links terms at different levels of complexity in facets 9, 8 and 4. Thus 8jd-u is 'clay fraction of illuvial horizon' and 4ni-rb is 'lime requirements'. The ordinal scale of the symbols used is as follows: stroke, capitals, numerals, hyphen, small letters.

9	Kinds, classification
b	Ortho-eluvium
bb	Colluvium
bd	Skeletal
c	Primary

cb	Secondary
cd	Residual
cf	Transitional
d	Azonal
e	Intrazonal
eb	Rendzina
ed	Mountain
ef	Regosol
eh	Planosol
f	Zonal
g	By origin (subdivide by a Rock or Mineral schedule)
h	By climate
hb	Arctic
hd	Temperature
i	Subtropical and tropical
ib	Laterite
id	Reddish brown
if	Yellowish brown
ih	Groundwater
ij	Red earth
im	Non-laterised
iq	Red loam
ir	Terra rossa
it	Regur
j	Humid
jm	Arid
k	Chernozem
kb	Degraded chernozem
kd	Black turf
kf	Smonitza
kh	Chestnut
kj	Reddish chestnut
km	Noncalcic brown
m	By physiography
mb	Desert
md	Red
mf	Grey
n	Prairie
nb	Reddish
nd	Brown steppe
nf	Reddish brown steppe
nh	High-veld prairie
p	Forest
pb	Rain-forest
pd	Swamp
pf	Lake
q	By constitution
qb	Allitic
qd	Siallitic
qf	Ferrallitic
qh	Lateritic
qj	Pedalfers
qm	Pedocals
qp	Calcareous

r	Peat soils
rd	Highmoor
rf	Transitional moor
rh	Lowmoor
rj	Bog
rm	Half bog
s	Podzols
sb	Brown
sd	Grey brown
sf	Grey wooded
sh	Groundwater
sj	Red
sm	Yellow
sn	Brown earth
sq	Gley
ss	Wet meadow
su	Alpine meadow
sw	Humic
t	Saline and alkaline
tb	Solonchak
td	Solonetz
tf	Solod
th	Szik
tj	Mallee
tm	By texture
u	Sand
v	Loamy sand
vn	Sandy loam
w	Loam
wn	Sandy clay loam
x	Clay loam
xn	Sandy clay
y	Clay
8	Structure
b	Single-grain
c	Granular
d	Crumb
db	Tilth
dd	Texture
e	Mature
eb	Immature
ed	Truncated
f	Profile
fb	Texture
fd	Mineral
ff	Organic
fh	Structure
i	Strata
id	Topsoil
if	Subsoil
ih	Ploughsole
ij	Capillary fringe
im	Water table
in	Rhizosphere

j	Horizon
jb	Eluvial
jd	Illuvial
jf	Recurrence
jh	A
ji	B
jk	C
jm	etc.
k	Pans
kb	Clay
kd	Hard
kf	Iron
kh	Lime
kj	Ortstein
km	Orterde
kp	Nazzaz
kz	Aggregates
m	Crumb
mb	Granule
md	Concretion
mf	Pea iron
mh	Puppet
n	Particles
p	Stone
q	Gravel
r	Sand
rb	Coarse
rd	Fine
s	Fine earth
t	Silt
u	Clay
v	Clay colloid
7	Constituents
b	Chemical
c	Inorganic, clay complex
cb	Mineral reserve
d	Nutrient
db	Availability
e	Trace elements
em	Silica/sesquioxide ratio
f	Nitrogen
fb	Ammonia
fd	Nitrite
ff	Nitrate
fh	Phosphorus
g	Other (subdivide by appropriate Chemistry schedule)
h	Water
hb	Gravitational
hd	Film
hf	Capillary
hh	Soil solution
hj	Groundwater
i	Organic, humus
j	Humic acid

jb Hymatomelanic acid
jd Brown humic acid
jf Gray humic acid
jh Fulvic acid
jj Humin
jm Carbon/nitrogen ratio
k Other (subdivide by appropriate Chemistry schedule)
kz Parent material
m Mineral, rocks
n Vegetation
nb Litter
nd Muck
nf Duff
nh Mor
nj Mull
nm Moder
np Gyttja
nr Peat
ns Other
p Biological organisms (subdivide by appropriate Biology schedule)
q Bacteria
r Fungi
s Actinomycetes
t Algae
u Protozoa
v Fauna
6 Properties
b Physical
c Mechanical composition
d Degree of aggregation
e Cohesion
f Consistence
g Plasticity
h Permeability
i Porosity
ib Capillary
id Non-capillary
m Other (subdivide by Physics schedule)
n Physico-chemical
p Moisture content
pb Water-holding capacity
pd Moisture equivalent
pf Hygroscopic coefficient
ph Sticky point
pj Plastic limit
pm Upper
pp Lower
pr Wilting point
pt Capillary potential
pv pF
q Exchange capacity
r pH
s Salinity
u Other (subdivide by appropriate Chemistry schedule)

v Biological
w Fertility
x Toxicity
z Other
5 Processes
b Formation
c Weathering
d Physical
e Chemical
f Solifluxion
g Profile development
h Humification
hb Anaerobic
hd Acid
hf Semi-arid
hh Humid
i Translocation
ib Infiltration, water movement
id Eluviation
j Drying, evaporation
k Chemical processes
m Fixation
mb Mineralisation
md Nitrification
mf Ammonification
mh Aminisation
mj Denitrification
n Adsorption
p Ion exchange
q Other (subdivide by appropriate Chemistry process schedule)
r Biological process (subdivide by a Biology process schedule)
s Leaching
t Exhaustion
u Erosion
ub Wind
ud Water
uf Accelerated
uh Gully
uj Rill
um Sheet
v Soil creep
w Landslide
4 Operations
b Survey
c Reclamation
d Drainage
db Mole
dd Impeded
df Tile
e Terracing
eb Bench
ed Broadbase
ef Step
eh Drainage

f	Irrigation
g	Tillage
gb	Ploughing
gd	Harrowing
gf	Levelling
gh	Hoeing
gj	Ridging
gm	Furrowing
gp	Rolling
h	Fallowing
j	Sterilisation
k	Conservation
m	Amendment
n	Amendment materials
nb	Inorganic
nc	Nitrogen
nd	Nitrate
ne	Ammonium
nf	Sodium
ng	Potassium
nh	Calcium
ni	Lime
nj	Gypsum
nk	Phosphate
nm	Superphosphate
nn	Sulphate
np	Sulphate of ammonia
nq	Compound fertilisers
nr	Industrial waste
ns	Basic slag
nz	Other (subdivided by a Chemical substance schedule)
p	Organic manure
pb	Vegetable
pc	Compost
pd	Animal
pe	Blood
pf	Bone meal
pg	Meat meal
ph	Farm
pi	Cattle
pj	Sheep
pk	Poultry
pm	Guano
pn	Municipal
q	Soil conditioners
r	Operations on amendment materials
rb	Requirements
rd	Collection
rf	Composting
rh	Grading
rk	Application
rn	Method
rq	Placement
rs	Quantity

ru Time
rw Storing
3 Experimental manipulations
b Physical
c Measurement
cb Units
d Methods
db Direct
dd Resonance
df Null
dh Compensation
dk Differential
dn Remote
dq Summation
ds Accuracy
du Errors
dw Corrections
e Recording
f Preparation
g Washing
h Drying
j Dissolving
k Extracting
m Chemical analysis (subdivide by appropriate Chemistry schedule)
q Biological manipulation
r Isolation
s Conservation
t Culturing
1 General
b Properties of properties and processes
c Intensity
d Stability
e Rate, Velocity
f Variation
g Limits
h Mechanism
i Cause
j Apparatus, equipment, device, instrument
jb Design
jd Construction
jf Specification
jh Testing
jk Calibration
jn Operation
jq Performance
js Maintenance
ju Standards
jz Operations on processes
k Initiation
kb Acceleration
kd Retardation
kf Prevention
m Control
n Automatic operation

p	Logical operations
q	Observation
qb	Experiment
r	Definition
s	Hypothesis
sb	Deduction
sz	Generalisation
t	Theory
tm	Law
u	Calculation
ub	Formula
ud	Model
v	Comparison
w	Classification
x	Nomenclature
y	Criticism

CONTAINER MANUFACTURE (D. J. FOSKETT)

Five facets are used: Products, Parts, Materials, Operations, and a set of Miscellaneous Common Sub-divisions which may be applicable to any part of the scheme. The facet indicators are capital letters. The notation is partly hierarchical (e.g at Cc), but ordinal whenever brevity seems to suggest it (e.g. at Cvm). Common sub-divisions are introduced by a point. Alphabetical breakdowns by name are enclosed in brackets—e.g. the Omnia cap is Cq (Omnia). The colon is used to subdivide by subject device—thus 'polythene coated bags' is BuDv : Btd. The ordinal scale of the symbols used is: point, colon, capitals, bracket, small letters. Abbreviated schedules follow.

B	PRODUCTS
b	Metal containers, cans
bb	Open top cans
bv	Flattened
bx	Self-heating
by	Lined
c	Sizes
cd	Fruit
d	Fish
e	Meats
f	Pet foods
h	Milk
hb	Condensed
hc	Evaporated
hd	Dried
hg	Pie can, tapered
hj	Cheese
hm	Jam
j	Beer
jm	Baby food

m	General line cans
md	Built up
me	Collar
ml	Extrusions
mp	Collapsible tubes
p	Aerosols
r	Non-metallic containers
rb	Composites
s	Cartons, folding boxes
sz	Bottles, phials
t	Flexible packages
td	Bags
tt	Tubing
tu	Tubes
tv	Collapsible tubes
tw	Squeeze bottles
ty	Film
u	Polythene film
v	Laminates
vc	Extrusion
vk	Wax
vl	Adhesion
w	Plastic articles
z	Miscellaneous (divide alphabetically) e.g. Drums

C		PARTS (abbreviated)
	b	Cylinders, bodies
	c	Ends, lids
	cd	Lever lid
	ce	Plug lid
	cg	Slip lid
	cj	Hinged lid
	cp	Easy opening
	cr	Tear strip
	cx	Captive
	d	Cones
	e	Rings
	f	Diaphragms
	g	Collars
	h	Valves
	j	Beads
	etc.	
	v	Joints, seams
	vb	End, double seam
	vc	Side
	ve	Lap
	vg	Lock and lap
	vh	Corner
	vld	Dry
	vle	Soldered
	vlf	Cemented
	vm	Notches
	vn	Double sliver

vp	Western
vr	Hinge, tongue and wire
vt	Tab and slot
w	Surfaces

D	MATERIALS
b	Metals
bc	Lacquered
bg	Untinned tinplate base
bw	Oil layer
bx	Tin coating
by	Tin–iron alloy layer
c	Tinplate
d	Aluminium
g	Paper and board
k	Plastics
kc	Polythene
l	Film
m	Laminates
q	Cork
qf	Rubber
qg	Glass
r	Solders (arranged by composition)
rs	Gaskets
s	Lining compounds
sx	Adhesives, cements
v	Coatings
vb	Lacquers (arranged alphabetically)
vd	Sizes
vj	Inks
vk	Fluorescent
y	Solvents
yv	Additives, modifiers

F	OPERATIONS (considerably abbreviated)
b	Analysis
bp	Application
bpe	Electrostatic
br	Adhesion
bv	Pre-treatment
c	Coating, lacquering, decorating
cv	Post-coating
d	Printing
ds	Stoving, drying, curing
dx	De-coating
f	Forming, tooling
fv	Winding, coiling
fy	Laminating
g	Assembling
j	Closing
jv	Proofing
jw	Cleaning
m	Testing
t	Coding

td	Labelling
tf	Packing
tg	Despatch
tm	Repair
tv	Safety
tx	Salvage
v	Storing

MISCELLANEOUS COMMON SUB-DIVISIONS

b	Research
c	Development
d	Committees
dd	Working parties
df	Meeting
dg	Societies
hb	Visits
hh	Courses
hj	Surveys
hk	Information
hr	Reports
k	Apparatus
kc	Instruments
l	Machines
lb	Whole
lc	Magazine
ld	Feed
lf	Conveyor
lh	Ejector
lj	Transmission
lk	Control
ll	Holder, chuck
lm	Tool
ln	Die
ly	Safety device
m	Measurement
mc	Properties
md	Dimensions
ms	Specifications
mv	Standards
n	Faults
q	Quality control
qs	Statistics
s	Supply
sb	Substitutes
v	Safety

APPENDIX C

CATEGORIES

To construct a classification, the traditional starting point has been the 'universe' of knowledge, which we then divide into conventional main classes. The 'entities' (concrete and conceptual existents) in each main class each have many attributes, and some of these can serve as characteristics for the 'logical division' of the main class into sub-classes or groups. The groups so derived can then be ranked in a helpful order. Each group is itself a fresh 'universe' susceptible of division by another characteristic. Successive division by a train of characteristics gives rise to a chain of classes. Diagrammatically, the result is the familiar 'tree of knowledge'.

Ranganathan, as well as others who advocate a faceted or 'analytico-synthetic' classification, proffers a modification to this pattern. He still starts with the traditional main classes or their canonical divisions, but suggests that any such original universe can be divided by more than one train of characteristics, each train giving rise to one facet of the universe. The facets encountered in the field of knowledge are all manifestations of five fundamental categories, and in a compound subject the most helpful order of citation of these categories is *personality*, *matter*, *energy*, *space*, *time*.

Effectively, therefore, Ranganathan's starting point is the series of terms which together comprise the *personality* category. Examples of these have been listed in Appendix A. Ranganathan has not attempted a strict definition of *personalities*, though he has on occasion equated them with 'wholes', and wrote that 'they are unanalysable and have to be handled gently and as a whole'—they are 'ineffable'. His approach cannot be accepted as it stands, because of the vagueness of the postulated category—he has not clearly defined it, nor has he analysed its relation to other postulated categories. Ranganathan's categories are, in fact, simply 'assumed'. They are postulated because he finds them to be of considerable help in subject analysis. 'Perhaps the analysis can be done equally well by postulating other fundamental categories', he wrote. 'This is only one of the many possible postulates. It is proving helpful. That is all.'

A firmer theoretical basis is needed than this. Upon the starting point of our classification all else depends. Bliss believed that in the traditional main classes he had found a permanent, stable basis. Whewell and Dove built their schemes on a series of conceptual categories—'ideas' or 'object-nouns'. The nineteenth-century materialists began with the concrete natural groups of entities. These workers did not preface their schemes with an analysis of the conceptual categories they used, to justify their choice of starting point, nor has Ranganathan offered such an analysis. This problem must be faced if we are to have a firm theoretical foundation for classification.

CONCEPTUAL CATEGORIES

The use of categories—concepts of high generality and wide application employed in the interpretation of the world—though relatively novel in classification, is as old as logic. Aristotle prefaced his own logical writings with one on 'the categories'. He recognised ten such terms:

substance (man, dog, stone, gold, timber, house, etc.)
quantity (large, two feet long, etc.)
quality (blue, loud, virtuous, etc.)
relation (double, heavier, slave, loudest, etc.)
place (here, at school, Greece, etc.)
time (yesterday, 1955, etc.)
situation (horizontal, upside down, sitting, standing, etc.)
state (fever, shod, healthy, etc.)
action (running, cutting, etc.)
passion (defeat, being cut, etc.)

For Aristotle the categories were terms, one or other of which must be affirmed of any subject if we ask what it is. They were the main types of entity involved in the structure of reality, the kinds of being to be recognised, the forms of existence which are realised in the actual existence of all things whatsoever.

According to Aristotle the basic category was that of *substance*, for if we ask of any concrete individual, what essentially it *is*, we must answer in terms of a substance: man, horse, stone, etc. The other categories all presuppose substances: if there were no objects, there would be no lengths or colours; if there were no animals, there would be no fever; if there were no knives, there would be no cutting or being cut.

How do conceptual categories arise? All concepts are ultimately based upon perception, but perception is not a static, fixed mode of acquiring knowledge. It is permeated by past experience, and as our experience broadens and deepens, so the categories which we use to interpret experience themselves develop in number, complexity and subtlety. The developing series of categories which we use to interpret external reality is the product of a progressively more profound penetration of that reality. To trace this development in detail is very difficult. It has its origin in the *social* activity of men cooperating to produce their means of subsistence, which they must do to remain alive. But at least an indication can be given of the way in which, in the *individual*, a series of simple conceptual categories is evolved.

THE DEVELOPMENT OF KNOWLEDGE

The acquisition of knowledge is an active process. It is a concrete interaction between the human organism and its environment, in the course of which the environment is physically and objectively changed, and the organism is mentally and subjectively changed. It is important to stress this point at the outset. In studying the development of conceptual categories, it is not only mental activities such as 'distinguishing' that have to be taken into account. Our total activity, mental and physical, is involved.

The first step in knowledge is, of course, perception, impressions, as in a flash. Then, from the confused blur of impressions, something is distinguished. The persistence or recurrent association of certain features of the sensed world leads on to the concept of the *object* or *thing*. We learn to distinguish a multitude of things, related at first only by virtue of the fact that they are all different. When two such distinct objects, 'isolates', are observed in juxtaposition, in association, the activity of the mind develops beyond bare observation to comparison. It is recognised that the difference between two objects is not absolute. Some of the associated sensations are common to both. Some things are alike, others unlike. Further, by the observation of a number of objects comparison grows into classification. Things are grouped into classes, members of a species, and species are differentiated from genera.

Some isolates, though distinguishable, are inseparable. They can only be conceptually sundered—the analysis of a substance into its sensed *qualities* (size, shape, colour, texture, and so on). Others can be physically sundered—the division of a substance into its *parts*.

Thus the practical experience of human experiment generates two further categories.

Throughout this development, while the human organism has been active—at first mentally and then physically—the isolates studied have been conceived statically, as purely passive play-things of our analytic and synthetic operations. But physical manipulation quickly proves in practice that this is far from true. *Parts* are not all inert appendages. Some can be separated from their parent whole without losing their characteristic behaviour—as sand can be separated from salt. But some cannot: when isolated, they rapidly lose their previous characteristics—a severed hand is no longer a hand. As Hegel put it: 'An animal may be said to consist of bones, muscles, nerves, etc., but evidently we are here using the term *consist* in a very different sense from its use when we speak of a piece of granite as consisting of quartz, felspar and mica. The elements of granite are utterly indifferent to their combination: they could subsist as well without it. The different parts and members of an organic body on the contrary subsist only in their union: they cease to exist as such when they are separated from each other.' Recognition of this fact leads to the development of two further categories—*constituent* (as quartz) and *organ* (as muscle).

The category of *organ* is widely applicable. In its biological usage, exemplified above, we recognise that the biological 'thing', the organism, is a hierarchy of organic parts. The body consists of organ-systems; these are composed of organs; organs include various tissues; these are made up of cells; and cells themselves have organic parts such as the nucleus. It is only when we reach the chemical substances of the body that we come to *constituents* which can subsist outside it.

We observe things in association, and learn to compare them. But things also *exist* in association, they are related to one another in innumerable ways. When we alter one thing—even if only by moving it—its relations to other things are altered, and these are revealed by changes in the things themselves. This becomes very apparent when we remove an organ from an organism, and observe the changes which occur in the organism to repair or compensate for the damage. Physical manipulation thus leads to the concept of the universal *interrelation* of things. An organism is thus related to its parts, and its parts to one another. For example, in embryology the development of each part of the body is determined by its relation to the body as a whole, but within this relationship a particular stage of the development of one part may be induced by its relation to another part.

Organisms—and, indeed, things generally—are also related to other organisms and to other things. Observation and experiment leads us to realise that such relations are not all casual or intermittent, but that some are connected and persistent. Things are grouped in *systems* and communities of various kinds. Celestial bodies, for example, form such groups as the solar system, constellations, star clusters and galaxies. Plants and animals form communities. Men and women form societies. There is a hierarchy of systems, one within the other. Each member of a system, and each sub-system, forms relations with other members, with other sub-systems, and with the super-system itself.

Relations between parts, things and systems are not static. They are displayed in action, and it is again by physical manipulation that our knowledge is generated. When we bring together salt and water a *process* occurs—solution, the salt dissolving in the water. In this process the things in relation mutually display their *properties*—the salt its solubility, the water its solvent power.

Moreover, even bare observation reveals that by no means all things are static even when relatively isolated. It is true that some things change little with time—a stone or a crystal. But for living organisms movement and *internal change*, behaviour, is life. The process may be rhythmic (as the beating of a heart) or serial (the development of an animal from birth to death). It is indeed such internal dynamism that leads to external *interaction*, and once again a distinction is generated. In some interactions the things affect each other, but remain independently recognisable; in other cases the reactants lose their identity. Thus a proton projected through matter may be retarded or reflected—these are *effects*—but in certain circumstances it combines with another particle and itself disappears—a *reaction* occurs.

Things and their material relations, though they form the basis of scientific investigation, do not exhaust the subject-matter of science. The investigator consciously interferes with the phenomena of nature, and this interference itself gives rise to a conceptual category. Thus the formation of an emulsion in an emulsifier, or the separation of a precipitate in a centrifuge, is something more than a liquid–liquid or a solid–liquid interaction. The emulsification and precipitation processes are further brought into an *operational* relation with the investigator. *Experimental operations* include production, use, observation, measurement and control. *Mental operations* are methods of manipulating phenomena theoretically, and include study, hypothesis, law, postulate, definition, theory and classification.

To carry out his operations man uses *tools* (instruments, devices, mechanisms). Experimental tools are innumerable—hammers, meters, balances, filters, emulsifiers, centrifuges, and so on. Mental tools include laws of thought, logic and dialectics, and mathematics. Each such tool before use has to be made, and is therefore at the same time a *product*—a 'thing' in its own right.

ILLUSTRATION FROM COLON AND THE UDC

To illustrate further the nature of these categories, and to give evidence that they are in line with recent practice in classification, it is worth examining the facets used in Colon and the UDC. The following table lists the categories used in that part of Colon which deals with science and technology. Each line gives the class number and name of basic *personality*, followed by any facets which depend on that *personality*. The 'fundamental category' of each facet is symbolised in brackets (P2=second-level *personality*, E=*energy*, 2P=second-round *personality*, 2M=second-round *matter*, 2E=second-round *energy*, etc.), and its name in the given class then follows, sometimes with an example. At the end of the table, the common facets also applicable to these classes are listed.

Facets in Colon, classes B to M

Basic personalities

B1	Arithmetic
B13	Numbers [P2] Problem (e.g. Primality) [E] Method (e.g. Analytical)
B2	Algebra
B23	Equation [E] Problem (e.g. Solution)
B25	Algebraic form [P2] Degree [E] Transformation (e.g. Quadratic)
B3	Analysis
B33	Differential and integral equation [P2] Degree [P3] Order [E] Problem (e.g. Solution)
B37	Real variable [E] Problem (e.g. Theory)
B38	Complex variable [E] Problem (e.g. Theory)
B4	Other calculus
B5	Trigonometry
B6	Geometrical space (e.g. Plane) [E] Method (e.g. Algebraic)
B7	Mechanical system (e.g. Solid) [E] Problem (e.g. Motion)
B8	Physico-mathematics
B9	Celestial body [E] Problem (e.g. Chronology, Astrophysics)
C1	Fundamentals of physics
C2	State of matter [E] Property (e.g. Density)
C3	Sound wavelength [E] Problem (e.g. Propagation)
C4	Heat [E] Problem (e.g. Transfer)
C5	Light wavelength [E] Problem (e.g. Propagation)
C6	Electricity [E] Problem (e.g. Production)
C7	Magnetism [E] Problem (e.g. Intensity)
C8	Cosmic hypothesis
C9	Specials (e.g. Nucleus, Surface, High Pressure)
CA	Systems (e.g. Relativity)
D1	Building material [P2] Part [E] Engineering (e.g. Construction)
D2	Irrigation [P2] Work [E] Engineering

D3 Mining substance [P2] Part of mine [E] Mining
D411 Road [P2] Part [E] Engineering
D415 Rail road [P2] Part [E] Engineering
D416 Bridge [P2] Part [E] Engineering
D42 Waterway [P2] Part [E] Engineering
D43 Air way [P2] Part [E] Engineering
D5 Transport vehicle [P2] Part [E] Engineering
D62 Mechanical part [P2] Purpose of machine (e.g. Generating) [E] Engineering
D635 Hydraulic engine [P2] Purpose [P3] Part [E] Engineering
D638 Pneumatic engine [P2] Purpose [P3] Part [E] Engineering
D64 Heat engine [P2] Purpose [P3] Part [E] Engineering
D65 Electronic device [P2] Purpose [P3] Part [E] Engineering
D66 Electric power [P2] Problem (e.g. Generation) [P3] Part [E] Engineering
D7 Nuclear engine [P2] Purpose [P3] Part [E] Engineering
D8 Municipal and sanitary engineering [E] Engineering
E Chemical substance [P2] Function group [E] Problem (e.g. Preparation)
F Substance [E] Process and problem
G1 Organ [E] Problem (e.g. Physiology) [2M] Substance
G19 Special grouping of organisms [E] Problem (e.g. Ecology)
H1 Mineral substance [E] Problem (e.g. Crystallography)
H2 Rocks [E] Problem (e.g. Genesis)
H3 Geological structure (e.g. Joint)
H4 Geological process (e.g. Earthquake)
H5 Stratigraphic horizon (e.g. Cambrian)
H6 Palaeontology
H7 Mineral deposits [E] Problem
H8 Cosmic hypothesis
I Plant group [P2] Organ [E] Problem [2M] Substance
I95 Ecological group [E] Problem (e.g. Ecology)
J Use [P2] Part [P3] Crop [P4] Organ [E] Farming

Farming is subdivided thus:

J:1 Soil [2P] Nature [2E] Operation (e.g. Drainage)
J:2 Manure [2P] Nature [2E] Operation (e.g. Collection)
J:3 Propagation [2P] Part [2E] Operation (e.g. Cutting)
J:4 Disease [2P] Agent [2E] Problem (e.g. Etiology) [3M] Drug
J:5 Development [2P] Operation (e.g. Thinning) [2E] Time
J:7 Harvesting [2P] Part [2E] Operation (e.g. Drying)
J9 Specials (e.g. Soil-less)
JA Systems (e.g. Forestry)
K Animal group [E] Problem
λ Use [P2] Part [P3] Farm animal [E] Farming

Animal husbandry is subdivided thus:

λ:1 Feeding [2P] Food [2E] Operation (e.g. Preparation)
λ:2 and 3 Morphology and Physiology
λ:4, 5 and 6 Disease [2P] Agent [2E] Problem (e.g. Etiology) [3M] Drug
λ:7 Harvesting [2P] Produce (e.g. Milk) [2E] Operation
λ:8 Training
λ:91 Nomenclature, classification
L Organ or organ-system [E] Problem (e.g. Physiology)
Problem includes
L4 Disease [2P] Agent [2E] Handling [3M] Drug
L9 Specials (e.g. Child, Aviation)

LA	Systems (e.g. Homoeopathy)
M1	Book production process
M2	Journalism
M31	Cooking
M4	Smithy
M5	Carpentry
M6	Glass industry
M7	Textile material [E] Work (e.g. Spinning)

etc., etc.

Common facets

[S]	Geographical division [P2] Physiographic feature
[T]	Chronological division
[P]	Institution [P2] Organ, official [E] Problem (e.g. Function)
[E]	Institutional activity (e.g. Function)
[E]	Industrial activity (e.g. Plan)
[E]	Intellectual activity (e.g. Hypothesis)
[M]	Property (drawn from schedule)
	Common divisions of literary form

Examples of naturally occurring *things* are of course easy to find: Chemicals in E, Minerals and Rocks in H, Plants in I and J, Animals in K and λ. *Products* are equally obvious in D and M. *Parts* of mechanisms appear in D and *organs* in I, J and L. *Systems* of organisms occur at G19 and I95. Among the *attributes* can be noted Primality in B13, Degree in B25 and B33, Density in C2, and Intensity in C7. Internal *processes* include Propagation in C3 and C5, Physiology in λ and L, Genesis in H2, and Earthquake in H4. *Experimental operations* occur widely: Construction in D, Preparation in E, many Farming operations in J and λ, Handling in L, Spinning in M7, and Industrial activity (a common facet). *Mental operations* include Method and Solution in B, Transformation in B25, and Intellectual activity (a common facet).

There are other facets to be considered: (1) The *personality* facets in B are not readily allocated to any of the categories just given. A number, an equation, a variable, a plane, are in fact conceptual entities, not physical, i.e. they are *mental products*. (2) Mechanical systems in B7 and States of matter in C2 are kinds of *thing*, but Sound, Heat, Light, Electricity and Magnetism are not things, but *motions* of things, as will be discussed in more detail later on. (3) The Purpose facet in D6 and D7 is introduced as a means of subdividing the preceding *personality* into kinds: it is in fact a characteristic of division, not a category of existence. This is also true of Nature and Agent in J, λ and L. (4) The Functional groups of E are in fact *parts* of chemical substances. (5) Ecology in G19 and I95 concerns *relations* between organisms. (6) Crystallography in H1 deals with an *attribute*, the structure of minerals. (7) The structures of H3 and the horizons of H5 are *parts* of the earth's crust.

The Colon classification also forms basic classes known as 'systems' and 'specials'. Among the former, Relativity in CA and Homoeopathy in LA are 'schools of thought' in physics and medicine. They are in fact mental interpretations of these fields, *mental tools*. But the 'system' Forestry in JA is parallel to the 'specials' Nuclear Physics in C9, Soil-less Farming in J9 and Child Medicine in L9: these are Physics, Farming and Medicine within specific boundary conditions, applied to specific *personalities*—the nucleus, a soil-less culture medium, the child.

This survey indicates that the facets actually used in Colon can be readily allocated to the categories developed here. The value of the categories I have formulated is that they are much more explicit than those of Ranganathan. It is, for example, clearer to class Intensity as a *quality*, Propagation as a *process*, Construction as an *operation*, and Ecology as a *relation*, than to class them all as 'problems' or 'manifestations of energy'.

We can now consider how the categories appear in the Universal Decimal Classification. The UDC does not have a consistently 'faceted' structure, but by means of its 'special auxiliary numbers' it has marked off a considerable number of more or less homogeneous groups of terms. A fairly wide selection of such auxiliary groups occurring in science and technology is given in the following table.

Sub-divisions in UDC, classes 5 and 6

Special auxiliary numbers	
521.03	Astrophysics (e.g. Density)
521.04	Astrochemistry (e.g. Chemical constitution)
521.07	Astrobiology
528.02	Measuring methods
53.01	Theory
53.02	General laws
53.03	Production
53.04	Effects (e.g. Physical)
53.05	Observation and recording
53.06	Use
53.07	Apparatus for production or study (e.g. Model)
53.08	Measurement
53.09	Dependence (e.g. on Pressure)
53–1 to 9	Geometrical figure (e.g. Line, Surface) (used also for 532)
533.6.01	Aerodynamics
534–1	State of matter
534–6 to 8	Wavelength
534–9	Geometrical space
534.01	Type of vibration (e.g. Transverse)
535–1 to 8	Type of radiation (e.g. Visible)
535.012	Isotropy and anisotropy
536–1 to 4	Conditions of thermal change (e.g. Reversible)

537–7	Electrical potential (e.g. Low)
537–9	Electrical frequency (e.g. Low)
54.01	Natural state, occurrence
54.02	Composition, structure
54.03	Physical property
54.04	Chemical property
54.05	Preparation, etc.
54.07	Synthesis
54.08	Formula, measurement
54.09	Use
54–1	State of substance (e.g. Liquid)
541.12.01	Phase state (e.g. 2-phase)
541.12.03	Dependence (e.g. on Pressure)
541.18.04	Phase separation method (e.g. Flocculation)
541.18.05	Dispersion method
543–4	Reagent, indicator
546–3	Oxygen and/or hydrogen compound
546–8	Salt
546–9	Polymer
547–1	Inorganic derivative
547–2 to 8	Organic derivative
547–91	State of substance
547–99	Polymer
547.0	Structural detail
547.024	Radical
547.53–02 to 04	Position of substituents
548.1.02	Crystalline system
550.34.01	Theory
550.34.03	Equipment
550.34.04	Standards
550.34.05	Data
553.06	Origin of deposit (e.g. Magmatic)
553.7.031	Temperature
56(11)	Stratigraphic horizon
571(–09)	Historical epoch
571(11)	Prehistoric era
572.02	Environmental influence (e.g. Climate)
575–181	Dimensions
575–182	Shape
575–183	Weight
575–184	Number
575.061	External characteristics (e.g. Colour)
576.8.06	Pathogenic action
576.8.07	Diagnosis method
576.8.093	Technique (e.g. Culture)
576.8.094	Morphology
576.8.095	Physiology
576.8.097	Antigen and antibody
577.15.01	Type of enzyme
577.15.02	Mode of action
577.15.03	Dependence (e.g. on Pressure)
577.15.04 and 05	Influence factor
577.15.06	Production
577.15.07	Synthesis, etc.

577.15.08	Determination, use in analysis
578.08	Technique (e.g. Breeding)
581.01 to 07	Influencing factors (e.g. Heat, Organism)
581.08 to 09	Technique (e.g. Culture)
591.04	Physics and chemistry
591.05	Metabolism
591.08	Technique (e.g. Breeding)
591.4.068	Type of organ (e.g. Rudimentary)
61–00	Morbid lesions (e.g. Inflammation)
61–02	Etiology
61–05	Characteristics of patient (e.g. Age)
61–07	Diagnosis, signs and symptoms
61–08	Treatment
61–09	Scientific study of disease
611–012	Teratology
611–013	Embryology
611–018	Histology
611–019	Comparative anatomy
611–08	Technique
612.014.4	Influencing factor
612.015	Physiological chemistry
612.016	Mechanism of attack and defence
612.017	Immunity
612.018	Hormone
612.019	Comparative physiology
615–011	Physical and chemical property
615–012	Preparation
615–014	Other aspect (e.g. Purification)
615–015	Activity
615–03	Method of administration
615–092	Mode of action
615–099	Intoxication
615.01	Pharmacology
615.03	Clinical treatment
615.06	Complications
615.079	Poisoning
616–018	Histology
616–1	Prosthetic material
616–2	Model, cast
616–7	Instrument
620.1.05	Apparatus
620.1.08	Measurement
621–05	Use
621–1	Characteristics of machinery
621–2 and 3	Part
621–4	Shape
621–5	Operation
621–6	Fuel
621–7	Maintenance, etc.
621–8	Power
621–9	Processing characteristics
621.01	Theory
621.3.02	Type of current
621.3.03 and 04	Part

621.3.05	Transmission system
621.3.06	Switching
621.3.07	Regulation
621.3.08	Measurement
621.3.09	Transmission properties
621.43–4	Fuel feed
621.43.03 to 06	Equipment
621.6.02 and 03	Displaced fluid
621.6.05	Use
621.7.01	Working
621.7.04	Treatment
621.7.06	Machinery
621.7.07	Tools
621.7.08	Testing
621.7.09	Finishing
621.73.04	Techniques
621.74.03	Object cast
621.744.4.06	Part of machine
621.774.03	Shape of pipe
621.783.03	Type of furnace
621.791.01	Weldability
621.791.02	Preparation
621.791.03	Equipment
621.791.04	Fluxes
621.791.05	Joints
621.8.02	Transformation of motion
621.8.03	Energy form
621.85.05	Part and arrangement of parts
621.882.082	System of screw-thread
621.892.09	Use of lubricant
621.9.02	Special tool
621.9.03	Substance worked on
621.9.05	Physical state
621.9.06	Machinery
622.01	Theory
622.03	Character of bed
622.04	Excavation method
622.24.05	Equipment
622.7.05	Treatment of product
622.7.06	Treatment of separating medium
622.7.07	Treatment of separated products
623.4.03 to 06	Part
623.43.08	Source of traction power
624.01	Building material
624.02	Part of structure
624.03	Arrangement of structure
624.04	Examination and calculation
624.05	Construction operation
624.07	Structural element
624.21.09	Part of bridge
625.03	Force and movement
625.06 and 07	Material
625.08	Equipment
625.2.01	Part of vehicle

627.8.01	Material
627.8.03	Conditions
627.8.04	Statics
627.8.05	Methods
627.8.07	Structural elements
628.9.0	Light source
629.1–2	Part of vehicle
629.1–4	Use
629.1–5	Operation
629.1–6	Fuel
629.1–7	Maintenance, etc.
629.1–8	Power
629.1.01	Part of vehicle
629.1.02 and 03	Mode of action of propulsion
629.1.04 and 06	Equipment of vehicle
629.7.01	Vehicle characteristics
629.7.02	Structure
629.7.03	Propulsion
629.7.04	Layout
629.7.05	Guidance
629.7.06	Auxiliary systems
629.7.07	Operation
629.7.08	Maintenance
636.08	Technique (e.g. Breeding)
639.1.0	Technique of hunting (also 639.2.0 Fishing)
64.02	Purpose of housekeeping
65.012	Administration
654.0	Organisation (also 656.0)
658.3.04 and 05	Influence factor
66–91	Physical state
66–93	Medium
66–94	Rate
66–95	Complexity
66–97	Temperature characteristics
66–98	Pressure characteristics
669.018	Properties
669.04	Heat processes
676.1.02	Operation (e.g. Boiling)
676.1.03 and 04	Material
676.1.05	Equipment
676.1.06	Product
678–1	Property
678–2	Part
678–4	Shape
678–9	Derivative
678.01	Property
678.019	Defect (e.g. Crack)
678.02	Operation (e.g. Forming)
678.03 and 04	Material
678.05	Equipment
678.06	Use
678.07	Physical nature of macromolecule (e.g. Thermosetting)
681.3.01	Principles
681.3.02	Configuration

681.3.04	Data representation
681.3.05	Data conversion
681.3.06	Programs
681.3.07	Storage

Common auxiliary sub-divisions

=	Language
(0)	Literary form
(1/9)	Locality
(–)	Boundaries
'0/9'	Time including '7' Occurrences in time
.001	Point of view, theoretical
.002	executive
.003	commercial
.004	use and working
.005	installation and outfit
.006	space and situation
.007	staff
.008	organisation
.008.0	form and structure
.009	social and moral
.009.0	form of relation

These 'facets' can be grouped into categories. Naturally occurring *things* appear as States of matter in 534–1, 54–1 and 541.12.01, as types of chemical substance in 543 to 547 and 678–9, as *constituents* of organisms at 576.8.097, 577.15.01 and 612.018. *Products* and tools are found at 53.07, 616–7, and widely in 62, and their *parts* are also often listed. As for *attributes*, we have Density in 521.03, Chemical constitution in 521.04, Property in 54.03, 54.04, 615–011 and 678, Temperature in 553.7, Shape in 575–182, 621–4, 621.774 and 678–4, Dimensions, Weight, Number and Colour in 575, Age in 61–05, and Crack in 678.019. *Processes* appear as Vibration, Radiation and Thermal change in 53, Physiology in 576.8, Embryology in 611, Inflammation in 61, Current in 621.3 and Intoxication at 615–099. *Interactions* between things occur under the headings of Effects, Dependence or Influencing factor in 53, 541.12, 572, 577.15, 581, 612 and 658, and also as Pathogenic action at 576.8.06. There are many groups of *operations* listed: Observation and Measurement in 53, Preparation and Synthesis in 54, Flocculation and Dispersion in 541.18, Technique in 576.8, 578, 581, 591, 611 and 63, Diagnosis in 576.8 and 61, Purification and Administration in 615, Maintenance at 621–5 and 621–7, Switching and Regulation in 621.3, Excavation and Treatment in 622, and Operation in 66, 676.1 and 678. *Mental operations* include Calculation in 624, and *mental constructs* such as Theory and Law occur in 53, Geometrical figure in 531 and 534 and Formula in 54.

The facets labelled Use and Purpose (53.06, 54.09, 621–05, 621.6.05, 621.892.09, 629.1–4, 64.02, 678.06) are all—like Purpose in Colon D—characteristics of division. Other facets are compound—e.g. Astrobiology is the study of living organisms on celestial bodies, and Physiological Chemistry is the study of the chemistry of organisms. The common 'points of view' .001 to .004 are largely mental and practical *operations;* .006 covers *products* specified by use; .008 refers to the *structure* and *operations* of an enterprise, and .007 to the *agents* of those operations (staff); lastly .009 is concerned with *relations* between people.

THE PRIMARY CATEGORY

The category which Ranganathan uses to differentiate homogeneous fields of knowledge, on which to base a general classification, is that of *personality*. For Aristotle, *substance* was primary, and in this he included things, their parts and systems. The notion of *personality* or *substance* is in no way esoteric—it is merely making explicit a category which is already widely used in practice.

It is a well-known fact that the central feature of Chemistry is the *chemical substance*, and the characteristic feature of chemical reaction is exchange of atoms between substances. It is equally simple to appreciate that *plants* and *animals* are 'substances' in the Aristotelean sense. It is perhaps not so easy at first sight to recognise the 'substances' of Physics. Heat, Light, Electricity—these are clearly forms of movement or energy, not 'things'. Yet the terms of Physics can be fairly readily related to various forms of *material substance*. Thus, dimensions, shape, mass, inertia, weight and density are all properties of material substance in general; attraction, rotation and vibration are similarly its motions, and energy, force, work, potential and frequency are measures of those motions. All these terms thus come together in the subject usually known as Mechanics.

On the other hand, the properties of hardness, elasticity, capillarity, viscosity, strength and so on do not apply to matter in general, but only to *masses*—solid, liquid or gas. Flow, fusion, deformation, friction, etc., are some of the motions of masses, with measures such as elastic moduli, melting point and the like. If *molecules* are interpreted to mean all sub-molar particles, the rest of Physics can be called the 'movement of molecules'. Sound and Heat are motions of molecules properly so-called, Light and Radiation are motions of photons, Electricity of electrons, Magnetism of dipoles. Such motions have measures such as wavelength, pitch, temperature and colour, and are subject to changes such as reflection, refraction.

diffusion, scattering, conduction and so on. Nuclear Physics is clearly based on the properties and motions of *nucleons*. There is nothing unorthodox about such a grouping of subjects, which is in fact followed by all the standard classifications, but the clear recognition that each section of Physics is based on a particular 'substance' is in accord with the modern development of the subject.

The notion of substance in such subjects as Astronomy and Geology is just as straightforward. The *celestial bodies* include stars, meteors, comets, the sun, the planets and the moon. *Systems* of celestial bodies are also forms of substance: the universe, galaxies, constellations, star clusters, the solar system. In Geology the *earth* is obviously the primary substance, followed by its *parts*—the lithosphere, the atmosphere—which are further subdivided into physiographic features (volcanoes, springs, seas, mountains, etc.) and still further into rocks, ores, minerals and so on. In Technology the various structures, materials, machines and devices made by man—and their parts—are 'substances' in the Aristotelean sense. The works of Art—what Thompson has called the 'realisation'—form the primary 'substances' in their field. In Sociology social groups at several levels can be recognised.

From all these examples it should be clear that the notion of 'substance'—though this use of the term may be unfamiliar to some—is quite straightforward once it has been explained, at any rate in the field of Science and Technology. An outline of the fields of knowledge so delimited has already been given on pp. 16–17.

As has been noted in Chapter 3, this list of fields approximately corresponds to a series of 'integrative levels' determined by the complexity of things. The use of this concept in developing the over-all structure of a scheme has been much explored by the Classification Research Group, particularly by D. J. Foskett and D. W. Austin. Their work has led to a listing of entities that may be given in outline as follows:

Physcial entities

Level	I	Fundamental particles
	II	Atoms, isotopes
	III	Molecules
	IV	Molecular assemblages, e.g. solid

Chemical entities

Level	I	Elements
	II	Compounds
	III	Complex compounds

Heterogenous non-living entities

Level	I	Minerals
	II	Rocks

	III	Physiographic features
	IV	Astronomical entities
Artefacts		
Level	I	Raw materials
	II	Processed raw materials
	III	Components
	IV	Finished articles
Biological entities		
Level	I	Viruses
	II	Organelles
	III	Cells
	IV	Tissues
	V	Organs
	VI	Systems, e.g. digestive systems
	VII	Organisms
	VIII	Communities, e.g. shoal, herd
Man		
Level	I	Individual
	II	Group
	III	Local community
	IV	National community
	V	International community
Mentefacts		
Level	I	Units, e.g. digit, note
	II	Words, numbers, bars, etc.
	III	Sentences, formulae, musical phrases, etc.
	IV	Paragraphs, themes, etc.
	V	Complete works, philosophical systems, etc.

Critical comment on this use of integrative levels has come from de Grolier and from Huckaby, and Roberts has specifically examined obscurities in the concept of *personality*.

BIBLIOGRAPHY

E. DE GROLIER. *A Study of General Categories applicable to Classification and Coding in Documentation*. Paris, 1962

D. J. FOSKETT. *Classification for a General Index Language*. London, 1970

D. J. FOSKETT and B. I. PALMER (eds.). *Sayers Memorial Volume*. London, 1961

S. A. S. HUCKABY. 'An enquiry into the theory of integrative levels as the basis for a generalised classification scheme'. *J. Docum.*, 1972, **28,** 97–106

LIBRARY ASSOCIATION. *Classification and Control*. London, 1970

N. ROBERTS. 'An examination of the personality concept'. *J. Lib.*, 1969, **1,** 131–48

APPENDIX D

THE CLASSIFICATION OF CHEMICAL SUBSTANCES

The historical study of general classifications, outlined in Appendix A, is a well-worked field. With the increasing growth and specialisation of knowledge, the problem of constructing special classifications has become very important. It is therefore timely if we look at special subjects, and study the history of their classification to win clues as to our treatment of them today.

One of the most difficult areas of knowledge to classify is that of chemical substances: there are so many of them, and they can be grouped in so many ways. We can form classes of substances containing the same chemical element, or the same radical, or having the same general structure (e.g. Polymers); we can group substances by chemical behaviour (e.g. Acids), by physical property (e.g. Elastomers) or by biological function (e.g. Hormones); we can classify them by use (e.g. Dyes) or by method of production (e.g. Fermentation products). What method of classification are we to choose?

The general library classifications of chemical substances are often based almost wholly on constitution. The Bibliographic Classification of Bliss is typical. Since the discovery of the Periodic Law by Mendeleev in 1869, the inorganic elements are divided into groups of similar elements, and the compounds of each are arranged according to composition. The organic substances are classed according to the radicals or 'functional groups' of elements which they contain.

- Minerals
- Inorganic
  - Non-metals
    - Hydrogen
    - Inert gases
    - Phosphorus group
    - Sulphur group
    - Halogens
- Metals
  - Alloys
  - Light Metals
    - Alkali
    - Alkaline earth
    - Zinc, etc.
    - Boron–aluminium family
    - Rare earths

- Metals
  - Heavy metals
    - Iron and steel
    - Cobalt, Nickel
    - Manganese
    - Chromium, etc.
    - Vanadium, etc.
    - Group IV (Tin, Lead)
  - Precious metals
    - Copper, etc.
    - Titanium, etc.
    - Platinum metals
- Radioactive elements
- Silicon, carbon
- Organic
  - Acyclic
    - Hydrocarbons
      - Saturated
      - Unsaturated
    - Alcohols
  - Carbonyls
    - Hydroxy-carbonyls
    - Acids
    - Sulphur
    - Nitrogen, Phosphorus
    - Organometallic
  - Isocyclic
    - Alicyclic
    - Terpenes
- Carbonyls
  - Isocyclic
    - Essential oils
    - Caoutchouc
    - Aromatic
      - Mononuclear
      - Condensed
      - Alcohols, etc.
  - Heterocyclic
  - Alkaloids
  - Lactones
  - Sterols
  - Carbohydrates

Is this emphasis on chemical constitution as a characteristic helpful? This is a question which must, ultimately, be put to the cotenmporary users of classification, but an historical survey of the problem will enable us to envisage other and perhaps more helpful ways of classifying chemical substances.

EARLY CLASSIFICATIONS

Even in ancient times, man had knowledge of an amazing variety of minerals. The use of metals such as copper, iron, silver, gold, lead, tin and mercury presupposes a knowledge of their ores and their oxides, as well as other derivatives such as the vitriols (sulphates). Clay, chalk, lime, gypsum and other useful materials such as talc, emery and asbestos; common salt, soda, potash and alum; sulphur, soot, charcoal and diamonds—these and many other inorganic materials were known to antiquity. Painters used ochre, cinnabar, orpiment, white lead, red lead and other colouring materials. The range of known organic substances was equally wide—petroleum, asphalt, amber, pine resins, tar-oils, numerous vegetable oils and animal fats, soap, starch and gums. Vinegar and other organic acids were obtained from fruits, and tannins from bark. Dyes from madder, indigo, saffron and other plants were widely used.

The progress of chemical techniques led in the Middle Ages to the recognition of more 'elementary' chemical substances—sulphuric and nitric acid, saltpetre, ammonium carbonate, lead and mercury nitrates, and several acetates. 'Combustible spirits of wine' (ethyl alcohol) was also recognised. The introduction of good distilling apparatus made it possible to prepare ethereal oils from plants—rose, violet, lavender, mint, aniseed, clove and many more. The flourishing metallurgical techniques of the sixteenth century added further inorganic substances such as the chlorides, nitrates and sulphates of gold, silver, iron, copper, tin, lead, mercury and antimony. Organic acids were isolated, such as succinic and benzoic. In the seventeenth century Rudolf Glauber made many outstanding additions to our knowledge of chemical substances. He prepared ammonium nitrate, potassium chlorate, permanganate, benzene, ethyl chloride, acetone, strychnine and other compounds.

The substances listed above represent only a small sample of the multitude of materials which the chemist has had to recognise, to describe, to name and to classify. As with plants and animals, there was at first no system in his procedures. But as his techniques developed, the chemist learnt which properties were significant in physically manipulating substances—and, hence, of value in conceptually classifying them.

The most obvious characteristic of the many inanimate substances found in nature is what we now call their *physical state.* The earth on which we walk is solid; the sea, rivers and rain are liquid; we are buffeted by the winds of the air, and warmed by the light of the sun from the sky or by the heat of the fire. Earth, water, air, and sky (or light, heat, fire)—these are the four 'principles', the 'elements', the types of matter recognised from time immemorial. In the Pyramid Texts of Ancient Egypt, the self-created sun-god Atum creates the world: 'Thou didst spit out what was Shū (air); thou didst sputter out what was Tefnūt (water).' These two then gave birth to Geb (earth) and Nūt (sky).

The four elements are found in the writings of the early Greek thinkers—particularly Empedokles—and the doctrine was considerably developed by the Hindu philosophers (*c.* A.D. 500) in whose writings we find inanimate substances classified as follows:

Váyù (air)
Tejas (fire, light)
Ap (water, organic acids, juices, milk, oils, etc.)
Earths
Rasas (mica, pyrites, bitumen, etc.)
Uparasas (sulphur, vitriol, etc.)
Sádhárana Rasas (sal-ammoniac, etc.)
Gems (diamond, emerald, etc.)
Metals
Salts (common salt, saltpetre, etc.)

Alkalies
Mineral earths

The earliest method of classifying chemical substances was therefore by physical state. Aristotle, however, considered that all such substances were mixtures of all four elements, and that the characteristic of division should be their *homogeneity*. Thus the Turkestan writer al-Farabi (*c*. A.D. 920), developing Aristotle's ideas, made the following classification:

- Chemical substances
  - With dissimilar parts
    - The parts consisting of dissimilar parts, e.g. meat, bone
    - The parts consisting of similar parts, e.g. minerals, stones
    - The parts not further partitioned, e.g. salt, gold, silver

Already in the Hindu sub-division of Earths we find that characteristics other than physical state and homogeneity are being used, even though unsystematically. The Uparasas, for example, are a group of *volatile* substances; Gems and Metals each have definite *physical properties;* Salts and Alkalies differ in taste and corrosive action. Such groupings appear also in the classification of the tenth-century Persian scientist al-Razi. He recognised the following chemical processes: primary substances were purified by means of distillation, decantation, filtration, roasting, digestion, amalgamation with mercury, washing, sublimation and calcination; substances were then brought either into a fusible state ('ceration') or into solution; solutions were combined, and the combinations fixed or 'coagulated' to form derivative products. The classification of substances to which these operations led al-Razi was as follows:

- Primary substances
  - Earthy
    - Spirits (mercury, sal-ammoniac, sulphur)
    - Fusible bodies (metals: gold, silver, copper, iron, tin, lead and probably zinc)
    - Stones (minerals, ores, mica, glass)
    - Vitriols (including alum)
    - Boraces
    - Salts (common salt, carbonates)
  - Vegetable (e.g. oils)
  - Animal (hair, blood, milk, etc.)
- Derivative (e.g. metal oxides and sulphides)

Here we find *behaviour towards heat* used to distinguish between the volatile Spirits, the fusible Metals and the inert Stones, and *chemical behaviour* used to differentiate between ácid Vitriols, alkaline Boraces and neutral Salts. Vegetable and animal products are introduced as well as mineral Earths, and the new idea—derived from practical experiment—that oxides and sulphides were compound Derivatives of simpler Earths.

RENAISSANCE CLASSIFICATIONS

The medieval alchemists clung to the conception of the four elements, though al-Razi (a critic of alchemy and a very experienced experimenter) had abandoned it. It was not until the developments in mining and metallurgy of the sixteenth century that any improvement was made in the classification of chemical substances. Many of the inorganic chemicals obtained by metallurgists began to be used by medical chemists such as Theophrastus Bombastus von Hohenheim, Paracelsus (1493–1542). In a text attributed to him, *Concerning the Nature of Things*, Paracelsus listed 18 species of corporeal things, each species having its own 'life' or essential characteristics: (1) metals, (2) mercury, (3) sulphurous (combustible) materials, (4) salts, (5) gems, (6) pearls, (7) the loadstone, (8) flints, (9) talc, cobalt, zinc, granites and other stones, (10) arsenicals, orpiment, realgar, etc., (11) excrement, (12) aromatic substances, (13) sweet substances, (14) resins, (15) herbs, roots and fruits, (16) wood, (17) bones, (18) flesh and blood. Paracelsus more or less clearly stated that earth, water, air and fire were not elementary substances, but 'fields' in which bodies exist.

Georg Agricola, in *De natura fossilium* (Basle, 1546), made the first scientific attempt to classify minerals, using the characteristics of fusibility, solubility, colour, odour and taste:

- Minerals
  - Simple
    - Earths (may be kneaded when moistened)
    - Solidified juices (liquefy when moistened)
      - Meagre
        - Liquid plus earth (salt, soda)
        - Liquid plus metal (rust)
        - Liquid plus mineral compound (vitriol, alum)
      - Unctuous (e.g. sulphur, bitumen)
    - Stones (e.g. gems, marble, rocks)
    - Metals (gold, silver, copper, iron, tin, lead, mercury, bismuth, antimony)
  - Compounds of simple minerals

In this, Agricola continued al-Razi's distinction between Simple and Compound substances. His first characteristic of division of the former was again *behaviour towards heat*, and Solidified juices were also subdivided according to whether they melted readily (Unctuous) or not. The Meagre juices, however, were divided according to the nature of the substances from which they were produced—for example, rust came from the action of liquid water on metallic iron.

Agricola dealt only with minerals. One of the first European chemists to attempt a classification of all the known types of chemical substance was J. J. Becher in his *Epistolae chymicae* (Amsterdam 1673). He recognised (1) 'minerae'—ores; (2) metals —in which he includes only certain 'natural' metals and mercury; (3) 'mineralia'—including bismuth, zinc, cobalt, antimony and magnesia; (4) 'salts'; (5) 'decomposita'—a mixture of all kinds of inorganic materials; (6) earths—mostly oxides; (7) 'distillata'—mineral acids, ammonia and spirits of wine; (8) oils; (9) 'limi'—various minerals; and (10) 'compositiones'—secret preparations.

A much more satisfactory classification of substances other than metals was provided early in the eighteenth century by G. Roth, in his *De salibus metallicis* (Halle, 1708). He relied heavily on *behaviour towards heat* to characterise substances, but recognised subtler variations of behaviour. The sulphurous substances, for example, were all combustible, but varied in volatility. The more volatile themselves might be divided into those which could be distilled and those which were very combustible. Roth also introduced new characteristics—the *chemical* distinction between alkali, acid and salt was developed, Combustible spirits were divided according to *mode of production* and Earths according to *solubility*. His classification is as follows:

- Saline substances
  - Alkaline
    - Fixed (soda, potash)
    - Volatile (ammonia)
  - Acid
- Unchanged (mineral acids)
    - Sweetened (ether, etc.)
- Salt-like substances
  - Salts
    - Fixed (common salt)
    - Volatile (e.g. organic acid salt)
  - Heavy metal salts
- Sulphurous (combustible) substances
  - Liquid, volatile
    - Distilled oils
    - Combustible spirits
      - Prepared by fermentation
      - Prepared by extraction

Less volatile
- Vegetable essences
- Expressed oils
- Solids (resins, sulphur, phosphorus)

Earths
- Fixed (e.g. burnt lime)
- Partly fixed (e.g. insoluble precipitates)

It can be clearly seen how the techniques of preparation—volatilisation, distillation, extraction, expression and so on—form the basis of Roth's classification. He grouped together those materials which required similar methods of handling. Since physical properties do reflect chemical composition, some of his classes do coincide with modern groups based on constitution—e.g. the alkalies and the mineral acids. In other cases, such as the earths, similar physical properties do not always imply similar constitutions. Roth clearly recognised acids and alkalies to be distinct from the other 'salt-like substances', though he still called them 'saline'. In 1744–54 G. F. Rouelle finally excluded them from the class of salts, defining the latter as the product of the action of an acid on a base.

THE CHEMICAL REVOLUTION

Though chemists had by the eighteenth century begun to put their substances into some sort of classified order, they had not yet created a systematic nomenclature whereby the position in that order of a substance was revealed by its name. Indeed, names—often based on superficial physical properties—sometimes led to grossly mistaken groupings. For example, the 'butters' (chlorides) of antimony, arsenic and so on were classed along with ordinary butter; 'oil of vitriol' (sulphuric acid) with olive oil; and 'spirit of salt' (hydrochloric acid) with 'spirit of wine' (alcohol). The names 'liver of sulphur' (impure potassium sulphide) and 'cream of tartar' (potassium hydrogen tartrate) were equally unhelpful. Substances were often named after their discoverers—Glauber's salt, alkahest of Van Helmont, Lilium Paracelsi, Boyle's fuming liquor, or pearly matter of Kerkringius. There was even a 'philosophical wood' (sublimed zinc oxide).

The new ideas introduced into chemistry by Antoine Lavoisier found expression in a sweeping reform of nomenclature. Lavoisier's prime contribution was to show that 'calxes'—the products of combustion of metals—were compound, in the sense used by al-Razi. About 55 'simple bodies' were recognised—many of them those that we now know as 'elements'. A new system of nomenclature was constructed, on the principles that a name should (a) indicate a substance, (b) define it, (c) recall its constituent parts, (d) classify it according to its composition and (e) indicate the relative proportions of its constituents.

As a result of Lavoisier's work, one of his collaborators, A. F. Fourcroy, was able to provide a much more systematic classification of chemical substances in his *Système des connaissances chimiques* (Paris, 1801). The broad division between Mineral, Vegetable and Animal, found already in al-Razi, was retained, and Minerals were divided into three groups by a composite characteristic: Earths and stones (insoluble, incombustible), Saline substances (soluble, incombustible) and Combustible bodies. The Earths were divided by two new characteristics—behaviour when struck against steel and behaviour towards an acid. The Combustible bodies included the Metals, which were classed according to their mechanical behaviour (brittle or ductile), their reaction to acid, and their behaviour on combustion in air. Despite the reform of chemical nomenclature which gave substances names that recalled their constituent parts, in Fourcroy's table of 1802 constitution was still not used as a basis for classifying chemical substances. A few decades later, however, this characteristic was used more freely in classification, as shown by the arrangement of the article on 'Chemistry' in the seventh edition of the *Encyclopaedia Britannica* (1842):

- Inorganic bodies
  - Simple substances
    - Supporters of combustion (oxygen, halogens)
    - Acidifiable bases (hydrogen, nitrogen, carbon, boron, silicon, phosphorus, sulphur, selenium, arsenic, antimony, tellurium, chromium, vanadium, etc.)
    - Alkalifiable bases
      - Alkaline bases (potassium and calcium groups)
      - Earthy bases (aluminium, etc.)
      - Difficultly fused bases (iron, nickel, cobalt, manganese)
      - Easily fusible bases (zinc, cadmium, lead, tin, bismuth, copper, mercury, silver)
      - Noble metals (gold, platinum metals)
  - Primary compounds
    - Acids
      - Oxygen acids
        - With simple base (e.g. nitric, perchloric, carbonic, oxalic)
        - With a compound base
          - Containing oxygen, carbon and hydrogen (e.g. acetic, lactic, citric)
          - Fatty
            - Solid (e.g. stearic)
            - Liquid (e.g. oleic)
            - Not soapifiable (e.g. cholesteric)
          - Resinous (e.g. pinic, silvic)
          - Containing oxygen, carbon and nitrogen (e.g. uric)
          - Containing oxygen, hydrogen, carbon and nitrogen (e.g. aspartic)

Containing oxygen, hydrogen, carbon and sulphur (e.g. xanthic)
Halogen acids (e.g. hydrochloric)
Cyanogen acids (e.g. prussic)
Sulphur acids (hydrogen sulphide)
Alkalies (e.g. potash, lime, and other oxides)
Neutral compounds
Water
Spirit (e.g. alcohol)
Ethers
Volatile oils
Fixed oils
Drying oils (e.g. linseed)
Fat oils (e.g. olive)
Solid oils (e.g. beeswax)
Bitumens (e.g. petroleum, tar)
Secondary compounds
Oxygen acid salts
Halogen acid salts
Cyanogen acid salts
Sulphur acid salts
Organic bodies
Vegetable
Acids (already included under Inorganic)
Alkalies (e.g. morphine)
Neutral principles (sugar, starch, gum, lignin, tannin, colouring matters, camphor, etc.)
Animal (blood, bile, milk, etc.)

In this classification the Simple inorganic substances (the chemical elements) are subdivided according to the nature of their 'bases' (oxides)—acidic or alkaline. The Primary compounds are also divided by *chemical behaviour* into Acids, Alkalies and Neutral compounds, a mode of division which was also introduced for the Vegetable bodies. The nature of the *constituent* elements was used to subdivided the Acids, the Oxygen acids, and the Secondary salts, but the more traditional characteristic of *volatility* was used for classifying the Neutral primary compounds (spirits, ethers, oils, etc.).

This 1842 classification marks the last stage before the final triumph of *constitution* as the dominant characteristic for classifying chemical substances. I have already illustrated its dominance in the *Bibliographic classification*. More detailed schemes for organic chemicals (Beilstein, Buhle, Dyson) are all based on only three characteristics—constituent elements such as Sulphur, constituent radicals such as Carboxyl, and type of chemical bond (e.g. Double bond). It would seem, therefore, that the modern trend in the classification of chemical substances is to abandon the earlier characteristics of division and to concentrate almost entirely on *constitution*, using *origin* only to class substances of unknown constitution, e.g. Natural products. This is undoubtedly the case for the classification of substances in 'pure' chemistry, but a different pattern is obtained if we look at classifications of industrial chemicals. Three of these, all published in the years 1948–49, are considered below.

The first, UDC class 66, uses the characteristics of *constitution* (e.g. Sulphur) and *origin* (e.g. Animal oils), but it also reverts to *mode of production* (e.g. Photochemical) and *physical state* (e.g. Colloidal). Moreover, it introduces another characteristic which is not found in any of the preceding schedules—that of *use*. This is in fact the most widespread method of division, characterising many classes such as Adsorbents, Beverages, Foodstuffs, Dyes and so on.

The chemical engineering classification of W. L. Bolles lists its headings alphabetically. As well as using all the characteristics found in UDC class 66, Bolles introduced *physical behaviour* (e.g. Plastics, which are outside 66 in the UDC) and *chemical behaviour* (Acids and Bases). The *Report on the chemical industry* of the Association of British Chemical Manufacturers, prepared after studying a number of other classifications, uses every characteristic (except homogeneity and bond type) that has been mentioned in this survey, and introduces a new one, the *scale of production* (Heavy or Fine). Their classification is summarised below.

- Sulphuric acid
- Alkalis and chlorine
- Other inorganic acids and salts, including ammonia
- Industrial gases (oxygen, hydrogen, nitrogen, rare gases, acetylene, propane, butane, carbon dioxide)
- Fertilisers—nitrogen, phosphatic, potash
- Heavy organic chemicals
  - Aliphatic alcohols and derivatives (e.g., aldehydes, ketones), including solvents and plasticiers
  - Organic acids
  - Intermediates such as nitrobenzene, aniline, and others made on a large scale
  - Synthetic detergents
- Medicinal fine chemicals
- Other fine chemicals
- Dyes and intermediates
- Pigments
- Explosives
- Agricultural products other than fertilisers
- Synthetic resins and plastics
- Miscellaneous
  - Ion exchange materials
  - Water softeners
  - Synthetic tannins
  - Synthetic waxes
  - Catalysts
  - Gels
  - Silicones
- Coal tar distillation products

EMERGENCE OF NEW CHARACTERISTICS

The results of this historical survey can be summed up in the following table, showing characteristics of division used in successive schemes.

	Physical state	Physical behaviour	Chemical behaviour	Origin	Mode of production	Constituent elements	Constituent radicals	Bond type	Use	Scale of production
Pyramid texts	×									
Hindus, A.D. 500	×	×	×	×						
al-Razi, 950		×	×	×		×				
Agricola, 1546		×			×	×				
Roth, 1708		×	×		×					
Fourcroy, 1801		×	×	×		×				
Britannica, 1842		×	×	×		×				
Ranganathan, Bliss			×	×		×	×			
Beilstein, Buhle, Dyson				×		×	×	×		
UDC class 66	×			×	×	×			×	
Bolles	×	×	×	×	×	×			×	
ABCM	×	×	×	×	×	×	×		×	×

Starting with the single character used in the Pyramid texts, physical state, a succession of new characteristics has been introduced: physical behaviour, chemical behaviour, origin, mode of production, constituent elements, constituent radicals, bond type, use, and scale of production.

If we looked only at modern schedules for organic chemicals, it might seem that as new characteristics are introduced, the earlier ones are dropped, but an examination of classifications for industrial chemicals shows that this is not the case. In fact, we find that even 'pure' chemistry, after a phase in which almost exclusive reliance was placed on constitution as a character of arrangement, is now readmitting other characteristics. Current organic chemistry texts include chapters on chemicals grouped by physical behaviour (Plastics, Solvents), by origin (Petroleum derivatives) and by use (Dyes, Drugs). Taking the grouping of chemical substances as a whole, the continuing emergence of new characteristics does not prevent the continued use of the old.

It must not be supposed that the development of new characteristics, useful in systematising chemical substances, has now ceased.

History does not come to a halt, and indeed we already have new methods of systematisation being advocated. For Inorganic chemicals, the usual method of arrangement is by constituent *element*, but this is not the only useful characteristic. We can classify compounds by means of *physical behaviour*. The chemist Hückel has recently suggested that we should 'differentiate firstly between two large groups of substances, namely volatile and involatile. . . . Substances of chemically distinguishable character and, as we now know, completely different kinds of atomic linkage are thereby separated from one another. A further subdivision of the involatile substances is into those which, in spite of their low volatility and their high melting points, are more or less easily brought into solution and those which are practically insoluble. A special place is allotted to the alloys.' Hückel's classification—which he adopts as the basis of his book—is therefore as follows:

- Chemical substances
  - Non-metallic
    - Volatile
    - Involatile
      - Soluble
      - Insoluble
  - Metallic

As he points out, it is by no means new. The four elements of Empedokles represents an incomplete classification of substances according to physical property. The characteristics of volatility and solubility were already implicit in the Hindu classification, and explicit in that of Roth. What is new, as Hückel emphasises, is that we can now relate physical behaviour to type of chemical *bonding*. We can replace the above scheme by the following:

- Chemical substances
  - Homopolar (the volatile and some insoluble involatile substances)
  - Heteropolar (the soluble involatile and some insoluble involatile substances)
  - Metallic
  - Mixed bonding character

CONCLUSIONS

This survey—brief and incomplete as it is—has, I hope, displayed the problem of indexing chemical substances in a way that an examination of contemporary schemes alone would not have done. The historical method of analysis has emphasised that, for comprehensive coverage, it is necessary to provide for groupings of chemicals by a considerable number of characteristics. There is a continuing emergence of new characteristics, and most of the old retain their utility. The survey has not only revealed this general feature of the problem, but has also helped us to isolate what are the characteristics to be considered.

In classifying chemical substances, we are therefore faced with a complex problem of *cross-classification*—the application of a number of characteristics to class the same individual substance. Cross-classification is anathema to traditional logic, but it is a necessity in the detailed systematisation of any field of knowledge. Some of the methods by which it can be handled in bibliographic classification have been discussed in earlier chapters.

The advance of knowledge in librarianship, as in any other subject, depends on the interplay of facts and ideas. For facts on which to base our decisions and our systems, we must look at the current situation. For ideas, we can with equal profit consult the past—for the present generation has no monopoly of imagination. Neglect of the past leads to a narrowness of viewpoint, and often to a repetition of past mistakes. The historical method is a useful tool which librarianship must not overlook in building up its research tradition.

BIBLIOGRAPHY

As general reference see, for example, F. FERCHL and A. SUSSENGUTH. *History of Chemistry*. London, 1939; *Sourcebook in Chemistry*, 1400–1900. New York, 1952

ABCM. *Report on the Chemical Industry*. London, 1949

W. L. BOLLES. *Petroleum Refiner*, 1949, **28,** No. 5, 140–148

E. L. BUHLE and others. 'An Organic Chemistry Classification'. *J. Chem. Educ.*, 1946, **23,** 375–391

M. CLAGETT. *Isis*, 1930, **14,** 373–384

G. M. DYSON. *New Notation and Enumeration System for Organic Compounds*. London, 1947

H. FRANKFORT and others. *Before Philosophy*. Harmondsworth, 1949

W. HÜCKEL. *Structural Chemistry of Inorganic Compounds*. Vol. 1. Amsterdam, 1950

E. H. HUNTRESS. *Brief Introduction to the Use of Beilstein's Handbuch*. New York, 2nd ed., 1938

P. C. RAY. *History of Hindu Chemistry*. Vol. 2. Calcutta, 1909

H. E. STAPLETON and others. *Chemistry in Iraq and Persia in the Tenth Century* A.D. Calcutta, 1927

A. E. WAITE. *The Hermetical and Alchemical Writings of Paracelsus the Great*. Vol. 1. London, 1894

NAME AND SUBJECT INDEX

Bibliographical references are in italic type